Rock Mechanics in Civil and Environmental Engineering

Rock Mechanics in Civil and Environmental Engineering

Contributors

Hayder Mohammed Salim Al-Maamori, Mohamed Hesham El Naggar et al.

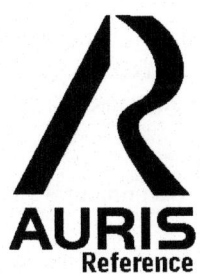

AURIS
Reference

www.aurisreference.com

Rock Mechanics in Civil and Environmental Engineering

Contributors: Hayder Mohammed Salim Al-Maamori, Mohamed Hesham El Naggar et al.

Published by Auris Reference Limited

www.aurisreference.com

United Kingdom

Copyright 2016

Printed in 2017 for Sale in the Indian Subcontinent

Rock Mechanics in Civil and Environmental Engineering

ISBN: 978-1-78154-839-4

British Library Cataloguing in Publication Data
A CIP record for this book is available from the British Library

Printed in the United Kingdom

Exclusively distributed by CBS Publishers & Distributors Pvt. Ltd.

Sales & Distribution Rights only for India, Pakistan, Bangladesh, Sri Lanka, Nepal and Bhutan. This book is not to be sold outside these territories.

Contents

List of Abbreviations

AE	Acoustic Emission
AMS	Anisotropy of Magnetic Susceptibility
BFEM	Base Force Element Method
DRM	Detrital Remanent Magnetization
EM	Electro-Magnetic
FEM	Finite Element Method
IRR	Infra-Red Radiation
JRC	Joint Roughness Coefficients
MTS	Mechanics Test Systems
NRM	Natural Remanent Magnetization
RAC	Recycled Aggregate Concrete
REV	Representative Elementary Volume
RMR	Rock Mass Rating
RQD	Rock Quality Designation
RSRM	Remote Sensing Rock Mechanics
SP	Self Potential
SPATE	Stress Pattern Analysis by Thermal Emission
SPT	Standard Penetration Test
TIR	Thermal Infra-Red
TSA	Thermo-Elastic Stress Analysis

List of Contributors

Hayder Mohammed Salim Al-Maamori
Department of Civil and Environmental Engineering, Western University, London, Canada

Mohamed Hesham El Naggar
Department of Civil and Environmental Engineering, Western University, London, Canada

Silvana Micic
Department of Civil and Environmental Engineering, Western University, London, Canada

Lixin Wu
Academy of Disaster Reduction & Emergency Management, Beijing Normal University, Beijing,, China
Institute for Geo-informatics & Digital Mine Research, Northeastern University, Shenyang, China

Shanjun Liu
Institute for Geo-informatics & Digital Mine Research, Northeastern University, Shenyang, China

Xuezai Pan
School of Mathematics, Nanjing Normal University, Taizhou College, Taizhou, China
Faculty of Science, Jiangsu University, Zhenjiang, China

Zhigang Feng
State Key Laboratory of Coal Resources and Safe Mining, China University of Mining and Technology, Beijing, China
Faculty of Science, Jiangsu University, Zhenjiang, China

Guoxing Dai
Faculty of Science, Jiangsu University, Zhenjiang, China

Hongguang Liu
Faculty of Civil Engineering and Mechanics, Jiangsu University, Zhenjiang, China

Andy A. Bery
Geophysics Section, School of Physics, Universiti Sains Malaysia, Penang, Malaysia

Rosli Saad
Geophysics Section, School of Physics, Universiti Sains Malaysia, Penang, Malaysia

Yifeng Chen
State Key Laboratory of Water Resources and Hydropower Engineering Science, Key Laboratory of Rock Mechanics in Hydraulic Structural Engineering, Wuhan University, P. R. China

Chuangbing Zhou
State Key Laboratory of Water Resources and Hydropower Engineering Science, Key Laboratory of Rock Mechanics in Hydraulic Structural Engineering, Wuhan University, P. R. China

Hu-Dan Tang
Key Laboratory of Ministry of Education for Geomechanics and Embankment Engineering, Institute of Safety and Disaster Prevention Engineering, Hohai University, Nanjing, Jiangsu 210098, China
School of Civil Engineering, Henan Polytechnic University, Jiaozuo, Henan 454000, China

Zhen-De Zhu
Key Laboratory of Ministry of Education for Geomechanics and Embankment Engineering, Institute of Safety and Disaster Prevention Engineering, Hohai University, Nanjing, Jiangsu 210098, China

Ming-Li Zhu
School of Energy Science and Engineering, Henan Polytechnic University, Jiaozuo, Henan 454000, China

Heng-Xing Lin
Water Conservancy Project Planning and Design Departments, Shanghai Investigation Design & Research Institute Co. Ltd., Shanghai 200434, China

Yijiang Peng
The Key Laboratory of Urban Security and Disaster Engineering, Ministry of Education, Beijing University of Technology, Beijing 100124, China

Qing Guo
The Key Laboratory of Urban Security and Disaster Engineering, Ministry of Education, Beijing University of Technology, Beijing 100124, China

Zhaofeng Zhang
The Key Laboratory of Urban Security and Disaster Engineering, Ministry of Education, Beijing University of Technology, Beijing 100124, China

Yanyan Shan
The Key Laboratory of Urban Security and Disaster Engineering, Ministry of Education, Beijing University of Technology, Beijing 100124, China

Yasuto Itoh
Graduate School of Science, Osaka Prefecture University, Osaka, Japan

Machiko Tamaki
Japan Oil Engineering Co. Ltd., Tokyo, Japan

Osamu Takano
JAPEX Research Center, Japan Petroleum Exploration Co. Ltd., Chiba, Japan

Irfan Celal Engin
Afyon Kocatepe University, Engineering Faculty, Department of Mining Engineering, Afyonkarahisar, Turkey

Preface

Rock mechanics is a theoretical and applied science of the mechanical behavior of rock and rock masses; compared to geology, it is that branch of mechanics concerned with the response of rock and rock masses to the force fields of their physical environment. Rock Mechanics in Civil and Environmental Engineering covers topics in the area of Rock Mechanics and related areas; covers recent developments in rock mechanics; shows how Rock Mechanics today has become more and more associated with, and indeed part of, construction, energy, and environmental engineering. First chapter presents a compilation of a number of in-situ stress measurements, strength and stiffness measurements, time-dependent deformation measurements, and some dynamic properties measurements of different rock formations in Southern Ontario and the neighbouring regions. Second chapter reveals on remote sensing rock mechanics and earthquake thermal infrared anomalies. In third chapter, we study roughness of center profile curve on rock fracture surfaces from statistical view. The main objective of fourth chapter is to determine the correlation in between seismic velocity values with engineering parameters such as N value, rock quality, friction angle, relative density, strength (force), consistency and velocity index. Beside than that, the correlation found also extent for good estimation which is important in engineering perspective especially for tropical region country. The main purpose of fifth chapter is to provide a theory for developing a stress-dependent hydraulic conductivity tensor for fractured rock masses. In sixth chapter, mechanical behavior of 3D crack propagation and coalescence is investigated in rock-like material under uniaxial compression. The purpose of seventh chapter is to survey the base forces element method on complementary energy principle for large-scale computing problems in rock engineering problems. Last chapter describes the microfabric of sedimentary rocks related to the tectonic regime and sedimentation processes in the mobile zone. It focuses on an attempt to apply magnetic properties to tectono-sedimentology.

Chapter 1

A COMPILATION OF THE GEO-MECHANICAL PROPERTIES OF ROCKS IN SOUTHERN ON-TARIO AND THE NEIGHBOURING REGIONS

Hayder Mohammed Salim Al-Maamori, Mohamed Hesham El Naggar, and Silvana Micic

Department of Civil and Environmental Engineering, Western University, London, Canada

ABSTRACT

The available measurements of the geo-mechanical properties of rocks in Southern Ontario and the neighbouring regions (New York, Ohio, Michigan, Indiana, Illinois, Wisconsin, and Minnesota) are summarized and presented. These measurements were compiled from available published data in the relevant literature and also from data that were collected from major underground projects in these regions. The compiled data are presented in three categories: measured in-situ stresses in different rock formations; calculated strength, stiffness and deformation including time-dependent deformation properties; and the measured dynamic properties of intact rock specimens from different rock formations in Southern Ontario and the neighbouring regions. The data presented in this paper can be used as a resource for preliminary evaluation of the geomechanical properties of the rocks in these regions. The presented geo-mechanical properties were generally obtained from in-situ measurements and from laboratory tests that were conducted on intact rock specimens from freshly excavated rock samples. Moreover, the time-dependent deformation properties of rocks in these regions were obtained from laboratory tests that were performed on intact rock specimens submerged in water. However, the influence of drilling fluids such as bentonite slurry and synthetic polymers solution, on the geo-mechanical properties of rocks is not evident and needs to be investigated.

INTRODUCTION

The first step in the design process of underground structures in rocks is to define the strength and deformation parameters of the rock unit in addition to the initial in-situ stresses that exist at a specific depth in the hosting rock unit. During the past few decades, extensive investigations of the initial in-situ stresses in rocks of Southern Ontario and the neighbouring regions (New York, Ohio, Michigan, Indiana, Illinois, Wisconsin, and Minnesota) and their strength and deformation properties including time-dependent deformation properties were carried out. The investigations revealed that the rocks of these regions are subjected to high initial horizontal in-situ stresses that are of great influence on the deformation behaviour of these rocks with time.

The deformation of the rocks with time is known as time-dependent deformation behaviour, which was manifested as different types of distress on the existing underground structures in Southern Ontario [1]. These distresses were observed in the form of cracks in the tunnels lining at the springline, invert heave, buckling of lining concrete of canal floors, bottom heaves in quarries; and long-term movement of walls of unsupported excavations [1]. In many cases, the resulting defects can cause severe damage on underground structures that requires costly remedial and maintenance works [1].

The time-dependent deformation behaviour of rocks in Southern Ontario was extensively investigated during the past decades [2] -[9]. Considering the osmosis and diffusion as a mechanism of swelling, these investigations were mainly based on measuring the swell deformation of intact rock specimens submerged in water with variable confining pressures and variable salinity of the ambient water. However, present-day tunnel drilling technologies such as micro-tunnelling and horizontal direction drilling involve fluids such as bentonite slurry and synthetic polymers solutions during the drilling process, which may influence the strength and time-dependent deformation behaviour of rock in the vicinity of the tunnel annulus. Bearing this in mind, it is quite indispensable to investigate the influence of these drilling fluids on the strength and time-dependent deformation behaviour of rocks in this region, and that research is ongoing at Western University. However, the research preceded with a comprehensive literature review which resulted in a compilation of available properties data obtained from tests performed on the intact rock exposed only to water.

Therefore, this paper presents a compilation of a number of in-situ stress measurements, strength and stiffness measurements, time-dependent deformation measurements, and some dynamic properties measurements of different rock formations in Southern Ontario and the neighbouring regions. The objective is that the presented data serve as initial source of information for any prospective study of the geo-mechanical properties of the rocks in these specified regions. Figure 1displays the locations of the sites from where data were compiled.

SUMMARY OF COMPILED MEASUREMENTS

In-Situ Horizontal Stresses

The available published values and directions of the in-situ horizontal stresses measured at different locations in Southern Ontario and the neighbouring regions were summarized and presented in Table 1. The presented data were compiled from sites where different measuring techniques were used to evaluate the in-situ stresses at variable depths and diversity of rock formations specifically in Southern Ontario and the surrounding regions (i.e. New York, Ohio, Michigan, Indiana, Illinois, Wisconsin, and Minnesota).

Table 1. *In-Situ* stresses in rocks.

Province/State/City	Project	Rock Formation	Rock Type	Depth (m)	Horizontal Minor stress (MPa)	Horizontal Major stress (MPa)	Direction of Major Horizontal Stress	Method Used	Source of Data
Ontario/Dufferin Creek	Outcrop in Duffin Creek, Ontario	–	Shale	9.1 - 15.2	6.9	–	–	USBM	[1]
Ontario/Elliot Lake	Mine in Elliot Lake, Ontario	–	Quartzite	390.0 - 415.0	21.4 - 44.1	–	–	–	[30]
			Diabase	256	15.2 - 41.4				
Ontario/Elliot Lake	Mine in Elliot Lake, Ontario	–	Sandstone/ Quartzite	204.8 - 701.0	17.24 - 22.06	20.69 - 36.54	East	OC	[29]
Ontario/Elliot Lake	Mine in Elliot Lake, Ontario	–	Sandstone/ Quartzite	427	24.13	35.37	–	USBM	[18]
Ontario/Kincardine	Bruce Nuclear Repository Site in Kincardine, Ontario	Cobourg	limestone	670	23	44.7	N 75°E	HF	[20]
Ontario/Mississauga	Heart Lake Tunnel in Mississauga, Ontario	Georgian Bay	Shale	6.0 - 18.2	0.86 - 6.32	1.25 - 9.5	N10° - 48 °E, N2° - 86°W	USBM	[2]
Ontario/Mississauga	Outcrop in Mississauga, Ontario	–	Shale	9.1 - 15.2	7.6	–	–	–	[1]
Ontario/Niagara Falls	SABNGS No3 in Niagara Falls, Ontario	Queenston	Shale	93.9 - 123.8	8.6 - 11.3	14.3 - 17.1	–	MSP	[16]
Ontario/Ottawa	Outcrop in Ottawa, Ontario	–	–	13.7	2.6	–	–	USBM	[31]
Ontario/Port Hope	Wesleyville Generating Station, Port Hope, Ontario	Trenton	Limestone	36.6	9.7	8.0 - 13.0	N 15°w	–	[1]
Ontario/Scarborough	Tunnel in Scarborough, Ontario	–	Shale	70.1	1.59	1.69	N 90°E	USBM	[31]
Ontario/Thorold	Thorold Tunnel In Thorold, Ontario	Gasport	Shaly limestone	18.3	6.63 - 12.7	8.14 - 14.69	N 60°E	USBM	[1] [7] [32]

Ontario/Thorold	Thorold Tunnel In Thorold, Ontario	Gasport	Dolomite	12.7 - 16.19	5.23 - 12.104	6.633 - 13.0	N27° - 88°W, N62°E	USBM	[1] [7] [32]
		Gasport	Dolomitic limestone	17.26	6.682 - 6.861	6.861 - 8.99	N60° - 76°E		
		Gasport	Fossiliferous limestone	19.82	6.647	13.833	N56°E		
		Gasport	Argillaceous limestone	24.7	6.848	10.513	N60°E		
		Gasport	Limestone with shaly interbeds	74.7 - 299.5	5.23 - 12.104	6.633 - 13.0	N27° - 88°W, N62°E		
Ontario/Thorold	Thorold Tunnel in Thorold, Ontario	Gasport member of Lockport and Decew formations	Dolomite	41.7 - 53.1	5.2 - 12.7	6.6 - 13	N27° - 88°W, N62°E	USBM	[24]
			Dolomitic limestone	56.6	5.2 - 6.6	6.8 - 9.03	N76°E		
			Shaly limestone	60.0 - 61.0	11.0 - 11.2	14.69	N58° - 60°E		
			Fossiliferous limestone	65	6.63	13.8	N56°E		
			Argillaceous limestone	81	6.83	10.5	N60°E		
Ontario/Thorold	Outcrop in Thorold, Ontario	–	Dolomite	12.7 - 15.5	5.21 - 12.07	9.03 - 12.07	N 27° - W, N 88°W	OC	[13]
			Dolomitic limestone	16.2 - 17.3	6.59 - 6.66	8.14 - 8.96	N 62°E, N 76°E		
Ontario/Thorold	Outcrop in Thorold, Ontario	–	Shaly limestone	18.3 - 18.6	11.03 - 11.17	14.69	N 60°E, N 58°E	OC	[13]
			Limestone	19.8 - 24.7	6.63 - 6.83	10.48 - 13.79	N 56°E, N 60°E		
Ontario/Wawa	Mine in Wawa, Ontario	–	Granite	341.4	40	60	–	–	[22]
Ontario/Wawa	Mine in Wawa, Ontario	–	Siderite	365.8	20.06 - 34.27	21.44 - 42.47	S 47° - 63°E	D	[28]
			Tuff	478.5	27.65 - 34.06	30.0 - 47.16	S 42° - 71°W		
			Meta - diorite	573	21.51	31.58	S 18°E		
			Chert	573	16.62 - 21.37	19.93 - 38.27	S 44°W, N 4°W		
Ontario/Wawa	Mine in Wawa, Ontario	–	–	332	27.9	–	–	D	[31]
Ontario/Darlington	Darlington Generation Station, Ontario	–	Ordovician limestone	228.0 - 300.0	10.5 - 11.3	17.2 - 19.6	N 70 E ± 7°	HF	[20]

Ontario/Toronto	Darlington Intake Tunnel, Toronto, Ontario	Whitby	Shaly limestone	74.7 - 299.5	5.8	9.3	N 63°E	_	[4]
Ontario/Toronto	Heart Lake Tunnel in Toronto	Georgian Bay	Shale	6.57 - 18.20	0.80 - 6.32	1.25 - 9.50	N 10° - 48°E, N 2 - 86°W	_	[3]
Ontario/North Bay	Outcrop in North Bay, Ontario	_	_	13.7	8.3	_	_	D	[31]
Ontario/Sudbury	Tunnel in Sudbury, Ontario	_	Jasperoid	45.7	44.82	51.71	_	_	[13] [31]
Quebec/Lake Beauchene	Tunnel in Lake Beauchene, Quebec	_	Gneiss W. Mica, Quartz	64	7.58	20	N 70°W	_	[13] [34]
Quebec/Churchhill	Cavern adit in Churchhill Falls, Quebec	_	Gneissic	305	11.72	13.79	_	OC	[35]
Quebec/James Bay	Mine in James Bay, Quebec	_	Monzonite/Syenite	121.9	5.48 - 11.24	8.14 - 20.69	N 0°E	D	[31]
Manitoba	Underground Research Laboratory in Manitoba	_	Granite	336.6 - 515	31.0 - 42.0	60.0 - 83.4	_	MSP	[26] [36]
Manitoba	Underground Research Laboratory in Manitoba	_	Granite	420	45	60	_	_	[23] [37]
Manitoba	Underground Research Laboratory in Manitoba	_	Granite	470.1 - 471.5	54.5 - 62.5	57.1 - 69.3	_	_	[38]
				579.5 - 670.8	56.9 - 76.0	61.0 - 76.7			
				745	46.8 - 51.8	57.9 - 61.5			
				836.9 - 851.3	56.2 - 78.3	62.6 - 85.7			
New York/Alma Township	Oil Field-Deep Boring in Alma Township, New York	_	Sandstone	502.9	10.17	15.69	N 77°E	HF	[19]
New York/Briarcliff Manor	Outcrop in Briarcliff Manor, New York	_	Gneiss	5.6 - 13.1	_1.48 - 3.62	_0.08 - 11.39	N 0°- 90°E, N64°- 74°W	OC	[13]
New York/Clarendon	Deep Borehole in Clarendon, New York	_	Sandstone/limestone	_	_	10.24	N 64°E	USBM	[31]
New York/Dale	Deep Boring in Dale, New York	_	Sandstone	_	11.89	18.61	_	HF	[13] [39]
New York/Niagara Gorge	Outcrop in Niagara Gorge, New York	_	Dolomite	0.2 - 6.7	_0.3 - 2.28	6.0 - 6.21	N34° - 55°E	OC	[13] [40]
New York/Nyack	Outcrop in Nyack, New York	_	Diabase	0.2 - 0.5	0.47	1.19	N 2°E	OC	[13] [41]

New York/Rochester	Sewer System in Rochester, New York	–	Dolomite	7.5 - 15.4	4.87 - 10.43	5.56 - 29.89	N10° - 86°E, N80° - 82°W	OC	[42]
New York/Somerset	Outcrop in Somerset, New York	–	Sandstone	8.5	3.17	4.41	N 15°W	OC	[13] [43] [44]
New York/Sterling	Outcrop in Sterling, New York	–	Sandstone	10.1 - 32.3	4.59 - 6.55	8.27 - 10.34	N22°- 90°W	OC	[13] [43] [44]
Illinois	Oil Field-Deep Boring in southern Illinois	–	Carbonate	99.1	2.41	7.76	N 62°E	OC	[17]
Michigan	Deep Boring in Gratiot Co., Michigan	–	Shale	5108	95	135	–	OC	[15]
			Sandstone	3660	67	90			
			Dolomite	3805	42	56			
Minnesota/Coldspring	Quarry in Coldspring, Minnesota	–	Granite	15	5.58	16.48	N 40°E	OC	[12]
Minnesota/Ely	Tunnel in Ely, Minnesota	–	Gabbro	305	10.3	16.5	–	OC	[12]
Minnesota/St. Cloud	Quarry in St. Cloud, Minnesota	–	Granite	–	10.58	15.1	N 50°E	D	[45]
Ohio	Boring in Ohio	–	Shale	10.3 - 18.6	4.69 - 32.41	5.58 - 38.13	N45° - 83°W N54° - 86°E	OC	[13]
Ohio/Barberton	Mine in Barberton, Ohio	–	Limestone	701	23.44	44.82	N 90°W	HF	[21]
Ohio/Falls Township	Oil Field-Deep Boring in Falls Township, Ohio	–	Sandstone	808	11.2	24.13	N 64°E	OC	[17]
Ohio/Hocking State Forest	Outcrop in Hocking State Forest, Ohio	–	Sandstone	0.9 - 1.2	0.37	0.63	N 61°E, N 83°E	OC	[14]
Wisconsin/Montello	Deep Boring in Montello, Wisconsin	–	Granite	75.0 - 188.1	6.2 - 8.2	14.0 - 20.0	N 63°E ± 20°	HF	[13] [46]

D: door stopper with South African CSIR strain cell; HF: hydro-fracturing technique; MSP: modified stress path method [16]; OC: over coring technique; USBM: the US bureau of mines deformation meter.

In general, one of the earliest attempts to measure the in situ stresses in rocks was made by Hast in the 1950's in Scandinavia as described in [11]. This attempt was followed by numerous studies that resulted in developing several methods to measure the in-situ stresses in different locations all over the world, many of which were in Southern Ontario. The most commonly methods to measure the initial horizontal in-situ stresses in rocks are: 1) the hydraulic fracturing (hydro-frac- turing test); 2) the over-coring technique with U.S. Bureau Mines probe (USBM); and 3) the under-coring technique with electrical strain gauges affixed in the borehole under consideration.

The hydraulic fracturing test consists essentially of sealing off a section of a borehole and injecting a fluid into the interval, inducing a fracture in the surrounding rock. The orientation of the resulting fracture and the pressures required to maintain the fracture are incorporated in an analysis to determine the in-situ stresses [12] [13]. The over-coring technique with (USBM) probe consists of drilling a hole to the required depth and then, from the bottom of this hole, a pilot hole of 38 mm diameter is drilled and the (USBM) probe is fixed in that hole. Then, the pilot hole is over-cored by employing a large diameter core bit to separate the rock core cylindercontaining the probe from in-situ. Later, the rock core cylinder is removed from the ground and tested in a hydraulic chamber to determine the modulus of elasticity and to calculate the in-situ horizontal stress using elastic theory relationships [13].

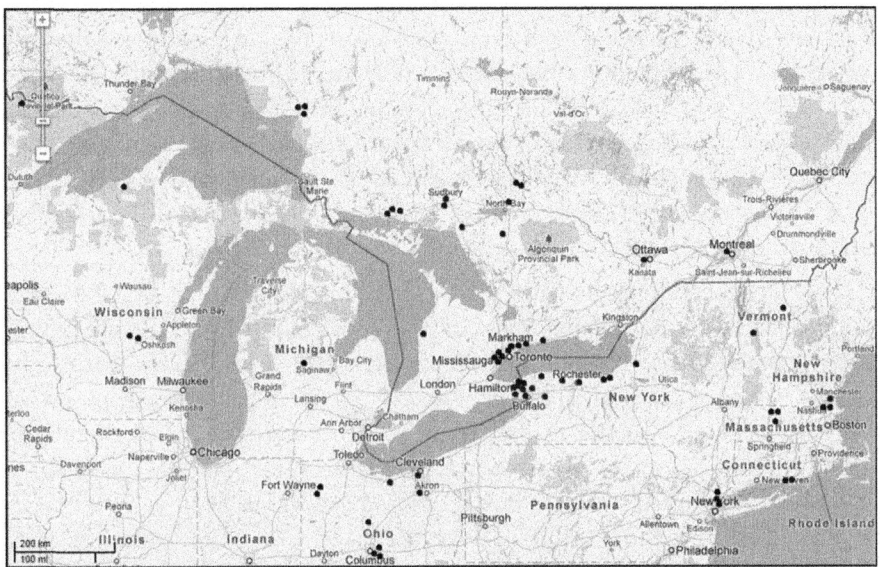

Figure 1. Locations of geo-mechanical data measurements [10].

The under-coring technique employs a package of electrical strain gauges, which is affixed to the base of the borehole. The waterproof electrical package and connections are sealed in a cylindrical form of plastic, and are affixed with quick setting epoxy at the bottom of the borehole. The deformation measurements of the borehole are taken before and after extending the core bit beyond the base of the borehole which under-cores the electrical strain gauges [13].

From the summarized data presented in Table 1, the value of the initial in-situ horizontal stress in rock formations of Southern Ontario and the neighbouring regions varies from a relatively small amount (<1 MPa) for sandstone in Ohio [13] [14] to a considerably high amount (>80 MPa) for sandstone in Michigan [15]. The high variation of the measured in-situ stress in rocks depends on the rock formation, type, depth and interbedded layers in the rock mass where stress measurements were taken. For example, the Georgian Bay shales in Toronto, Ontario possess an initial in-situ horizontal stress of a considerably high value of 1.25 - 9.5 MPa in the major horizontal stress direction and 0.86 - 6.32 MPa in the minor horizontal stress direction at depth of 6.0 - 18.2 m [2] [3]. The Queenston shale from the Niagara Falls area, Ontario, exhibits an initial in-situ horizontal stresses of 14.3 - 17.1 MPa in the major horizontal stress direction and 8.6 - 11.3 MPa in the minor horizontal stress direction at depth of 93.9 - 123.8 m [16]. In addition, shale in Ohio, at 10.3 - 18.6 m depth, possesses comparatively high in-situ horizontal stresses of 5.56 - 38.13 MPa and 4.69 - 32.41 MPa in the major and minor in-situ horizontal stress directions, respectively [13]. In the presented data, the highest measured in-situ horizontal stresses in shale of North America were recorded in Michigan, where the stress measurements were taken at overwhelming depths that exceeded 5100 m. The measured in-situ horizontal stresses in shale of Michigan at that depth were 135.0 MPa and 95.0 MPa in the major and minor in-situ horizontal stress directions, respectively [17].

On the other hand, sandstone of Elliot Lake, Ontario, at 427.0 m depth, exhibits an in-situ horizontal stress of 35.37 MPa and 24.13 MPa in the major and minor in-situ horizontal stress direction, respectively [18], while for similar depths in New York State the in-situ horizontal stresses in the sandstone were varying from 10.17 MPa in the minor in-situ horizontal stress direction to 15.69 MPa in the major in-situ horizontal stress direction [19]. In Michigan, the in-situ horizontal stresses were measured at 3660 m deep in the sandstone layer and were found as high as 90.0 MPa and 67.0 MPa in the major and minor in-situ horizontal stress directions, respectively [15].

The limestone in Kincardine, Ontario and the limestone in Barberton, Ohio exhibits considerably high in-situ horizontal stresses of 44.7 MPa and 23.0 MPa in the major and in the minor in-situ horizontal stress directions, respectively, at depths of around 700 m [20] [21]. Similarly, the measured in-situ horizontal stresses at 341 - 420 m depth in the granite layer in Wawa, Ontario and in Manitoba were as high as 60.0 MPa in the major in-situ horizontal stress direction and 40.0 MPa in the minor in-situ horizontal stress direction [13] [22] [23]. Although the in-situ vertical stresses from the overburden are not presented here in the compiled data, it could be perceived that in general, the rock formations in Ontario and neighbouring regions are subjected to a considerably high in-situ horizontal stresses.

Lo [1] analyzed natural geological features, such as: faulting; folding and buckling or pop-up of surface rock strata; distress in shallow and deep excavation, such as heaves in the Dufferin quarry in Milton; jamming of wheel pit, bending and buckling of steel beams structures of hydro-electric power plants; and crushing and spalling of arch and floor heave of the hydro tunnels in the Niagara area and Chippawa Canal in Ontario. Based on these analyses, it was suggested by Lo [1] that these observations were evidence of high in-situ horizontal stresses that resulted from the current movement of continental drift according to tectonic theory, and not due to the past overburden load during glaciation ages [1]. From the recorded in-situ stress measurements and the observation of natural phenomena, it was proposed that the belt of high horizontal stresses stretches from Rochester in New York State westward through Niagara Falls, turning northeast around Lake Ontario following the lake shore line and extending at least as far east as Wesleyville, Ontario [1].

The high in-situ horizontal stresses in rocks are a general phenomenon that exists in many regions in North America and the world. However, the rock formations in Southern Ontario and the neighbouring states, in specific, exhibit a considerably high in-situ horizontal stresses. These high in-situ horizontal stresses, after their relief, might be of significant influence on the time-dependent deformation characteristics of these rocks, which in turn might cause serious damages to the constructed underground structures.

Intact Rock Strength and Stiffness Properties

The values of the tensile strength, compressive strength, elastic (Young's) modulus and Poisson's ratio of different rock formations in Southern Ontario and the neighbouring regions are summarized and presented in Table 2. The presented data were compiled from available relevant literature.

Table 2. Intact rock strength and deformation properties.

Province/State/City	Project	Rock Formation	Rock Type	Depth/Elevation (m)	Tensile Strength (MPa)	Uniaxial Compressive Strength UCS (MPa)	Elastic Modulus E (GPa)	Poisson's Ratio v	Swelling Potential (%)	Source of Data
Ontario/Elliot Lake	Mine	–	Quartzite	390	–	31.0 - 44.1	80.0	–	–	[18]
Ontario/Elliot Lake	Mine	–	Sandstone/Quartzite	204.8 - 701.0	–	–	76.0	–	–	[29]
Ontario/Elliot Lake	Mine	–	Quartzite	390 - 415	–	–	80.0	–	–	[30]
		–	Diabase	256	–	–	93.0	–	–	

Group	Unit	Rock type							[8] [20]
Typical Values From Different Sites In Southern Ontario For The Bruce Nuclear Site	Lockport Goat Island	Dolostone	–	–	137.0 - 282.0	58.0 - 81.0	0.2 - 0.4	0.0 h
	Lockport Gasport	Shaly limestone	–	–	27.0 - 255.0	25.0 - 70.0	0.1 - 0.5	0.08 h
	De Cew	Dolostone/Mudstone	–	5	74.0 - 174.0	43.0 - 57.0	0.3 - 0.4	0.04 h
	Irondequoit	–	–	–	60.0 - 185.0	50.0 - 78.0	0.1 - 0.5	–
	Reynales	–	–	–	53.0 - 141.0	22.0 - 49.0	0.2 - 0.5	–
	Cabot Head	–	–	5.0 - 14.0	20.0 - 127.0	–	–	–
On-tario/Kin-cardine	Queenston	Shale	–	1.0 - 15.0	12.0 - 118.0	7.0 - 34.0	0.1 - 0.5	0.3 h
	Georgian Bay	Shale	–	–	3.0 - 206.0	1.0 - 58.0	0.1 - 0.5	0.15 h

Ontario/Kincardine — Typical Values From Different Sites In Southern Ontario For The Bruce Nuclear Site

Formation	Rock Type			22.0 - 140.0	10.0 - 67.0	0.1 - 0.6	[8] [20]
Cobourg	—	—	—	—	—	—	—
Lockport Eramosa	Dolostone	—	—	118.0	63.0	0.4	0.0 h
Rochester	Shale	—	—	85.0	23.0	—	0.07 h
Grimsby	Sandstone/Shale	—	—	25.0	8.0	—	0.27 h
Power Glen	Sandstone/Shale	—	—	26.0	9.0	—	0.17 h
Blue Mountain	Shale	—	—	27.0	2.0	—	0.15 h
Collingwood	Black shale	—	—	80.0	20.0	—	0.0 h
	Grey mudstone	—	—	58.0	10.0	—	0.15 h
Lindsay	Limestone with shaly interbeds	—	—	110.0	46.0	—	0.05 h
Verulam	Shaly limestone	—	—	23.0	57.0	—	0.05 h
Gull River	Limestone	—	—	143.0	63.0	—	0.0 h
Precambrian	Medium grained	—	—	190.0	60.0	—	0.0 h

Location	Site	Formation	Rock type							Ref
Ontario/ Kincardine	Typical values from different sites in Southern Ontario for the Bruce nuclear site	Granitic Gneiss	Coarse grained	–	–	140	46	–	0.0 h	[8] [20]
Ontario/ Kincardine	Typical values from different sites in Southern Ontario for the Bruce nuclear site	Amherstburg	Dolostone	–	–	33.0 - 113.0	8.0 - 40.0	–	–	[8] [20]
Ontario/ Kincardine	Typical values from different sites in Southern Ontario for the Bruce nuclear site	Amherstburg	Limestone	–	–	23.0 - 182.0	12.0 - 66.0	–	–	[8] [20]
Ontario/ Mississauga	Heart Lake tunnel	Georgian Bay	Shale	6.0 - 18.2	–	–	12.4	0.15	–	[2]
Ontario/ Niagara Falls	Sir Adam Beck Niagara generating station (SABNGS) No. 3	Queenston	Shale	95.64 - 114.33	–	–	–	–	0.22 - 0.34 h 0.37 - 0.54 v	[6] [8]
Southern Ontario	Different Sites In Southern Ontario	Rochester	Interbedded shale and Dolomite	–	–	20.0 - 40.0	20.0	–	–	[1]
Southern Ontario	Different Sites In Southern Ontario	Georgian Bay	Interbedded shale/ Siltstone/ Mudstone/ Limestone	–	–	30.0 - 190.0	20.0 - 40.0	–	–	[1]
Southern Ontario	Different Sites In Southern Ontario	Collingwood	Interbedded shale/Mudstone	–	–	20.0 - 70.0	7.0 - 20.0 v 14.0 - 35.0 h	–	–	[1]

Location	Site	Formation	Rock type							Ref.
Ontario/Sudbury	Tunnel	—	Jasperoid	45.7	—	—	83.0	—	—	[13] [33]
Ontario/Thorold	Outcrop		Dolomite	12.7 - 15.5	—	—	71.0 - 73.0	—	—	[13]
			Dolomitic limestone	16.2 - 17.3	—	—	73.0 - 74.0	—	—	
		—	Shaly limestone	18.3 - 18.6	—	—	43.0	—	—	
			limestone	19.8 - 24.7	—	—	55.0	—	—	
Ontario/Thorold	Thorold tunnel	Gasport member of Lockport/De Cew formations	Dolomite	12.7 - 53.1	—	—	71.0 - 73.0	0.27 - 0.3	—	[24]
			Dolomitic limestone	56.6	—	—	74.0	0.3	—	
			Shaly limestone	60.0 - 61.0	—	—	43.0	0.25	—	
			Fossiliferous limestone	65.0	—	—	55.0	0.3	—	
			Argillaceous limestone	81.0	—	—	55.0	0.3	—	
Ontario/Toronto	Darlington intake tunnel	Whitby	Shaly limestone	83.4	—	52.0 - 63.3 h	52.9 - 54.6 h	0.25 - 0.27 h	—	[25]
				84.4 - 84.7	—	87.6 - 88.2 v	39.6 - 43.6 v	0.34 - 0.37 v	—	[47]

On-tario/To-ronto	Domed stadium	Geor-gian Bay	Shale	19.8 - 26.3	–	11.2 - 17.2	2.2	0.3	–	[4]
On-tario/Wawa	Mine	–	Siderite	365.8	–	–	67.6 - 118.0	–	–	[28]
			Tuff	478.5	–	–	68.3 - 115.8	–	–	
			Meta-diorite	573.0	–	–	52.4 - 70.3	–	–	
			Chert	573.0	–	–	51.7 - 80.0	–	–	
South-ern On-tario	Research program for the National Research Council of Canada, different sites in Southern Ontario	Lock-port	Dolomitic limestone	157.0 - 168.0	–	180.0 h 200.0 v	76.0 h 67.0 v	0.14 - 0.33	0.02 h 0.01 v	[7]
			Gasport shaly lime-stone	–	–	105.0 h 120.0 v	44.0 h 27.0 v	–	0.08 h 0.08 v	
		Roches-ter	Shale	26.2 - 26.52	–	70.0 h	27.0 h	–	0.07 h 0.16 v	
		Geor-gian Bay	Shale	10.17 - 15.33	–	35.0 h	21.0 h	0.06 - 0.25	0.03 - 0.14 h 0.2 - 0.22 v	
		Colling-wood	Grey Mud-stone	17.0 - 24.64	–	35.0 h 60.0 v	23.0 h 10.0 v	0.2	0.15 h 0.45 v	
			Black shale	17.0 - 24.64	–	70.0 h 80.0 v	37.0 h 20.0 v	0.1 - 0.25	0.0 h 0.0 v	

Ref.	Region	Program	Formation	Rock type					
[7]	Southern Ontario	Research program for the National Research Council of Canada, different sites in Southern Ontario	Trenton-Black River	Limestone	12.9 - 35.5	—	130.0 h / 55.0 v	55.0 h	0.19 - 0.4 / 0.0 v	0.0 h
[7]			Trenton-Black River	Shaly limestone	12.9 - 35.5	75.0 v	100.0 h	57.0 h	—	0.06 h 0.09 v
[7]			Queenston	Shale	—	—	—	—	—	0.04 h 0.14 v
[5]	Southern Ontario	Research program for the National Research Council of Canada, different sites in Southern Ontario	Lockport	Shaly limestone (Gasport)	159.94 - 162.05	—	124.0 - 212.0 v / 27.0 - 115.0 h	25.3 - 61.2 v / 47.8 h	0.14 - 0.29 v / 0.24 h	—
[5]			Lockport	Fossiliferous limestone (Gasport)	159.23	—	152.0 v	59.1 v	0.24 v	—
[5]			Lockport	Fossiliferous limestone (Gasport)	157.33	—	102.0 h	75.9 h	0.33 h	—
[5]			Georgian Bay	Shale	15.33	—	35.0 v	5.5 v	0.13 v	—
[5]			Georgian Bay	Shale	10.17	—	41.0 h	12.1 h	0.06 - 0.25 h	—
[5]			Collingwood	Black Shale	22.76	—	80.0 v	20.4 v	0.18 v	—
[5]			Collingwood	Black Shale	18.49	—	25.0 - 72.0 h	14.8 - 38.0 h	0.10 - 0.15 h	—
[5]			Collingwood	Black Shale	16.99	—	21.0 i	13.4 i	0.09 i	—

Source	Formation	Rock type							Ref
Research program for the National Research Council of Canada, different sites in Southern Ontario	Collingwood	Grey Shale	23.34	—	58.0 v	9.8 v	0.2 v	—	[5]
	Trenton	Shaly limestone	23.28	—	32.0 - 35.0 h	19.7 - 26.0 h	0.09 - 0.15 h	—	
			12.93 - 26.06	—	84.0 - 129.0 h	53.4 - 60.5 h	0.19 - 0.39 h	—	
	Trenton	Limestone	35.41	—	75.0 v	54.8 v	0.35 v	—	
			35.48	—	133.0 h	54.8 h	0.24 - 0.4 h	—	
			35.53	—	91.0 i	45.7 i	0.35 i	—	
	Rochester	Shale	26.37	—	85.0 v	22.5 v	0.16 v	—	
			26.24 - 26.52	—	61.0 - 85.0 h	21.8 - 32.3 h	0.24 - 0.26 h	—	
			26.29	—	40.0 i	19.0 i	0.39 i	—	
	Lockport	Dolomite (Goat Island)	168.17	—	246.0 v	64.0 v	0.29 v	—	
			168.1	—	207.0 h	63.3 h	0.31 h	—	
	Lockport	Dolomitic limestone (Gasport)	165.15	—	208.0 v	57.7 v	0.32 v	—	
Research program for the National Research Council of Canada, different sites in Southern Ontario	Lockport	Dolomitic limestone (Gasport)	165.07	—	46.0 h	—	—	—	[5]
	Lockport	Dolomitic limestone/ Limestone (Gasport)	163.86	—	199.0 v	61.2 v	0.25 v	—	
			163.78	—	191.0 h	63.3 h	0.28 h	—	

									[3]
Southern Ontario	Thorold tunnel, wheel pits in the Canadian Niagara falls and Toronto power g.s., heart lake tunnel in Mississauga, intake tunnel of Darlington g.s., Scotia plaza in Mississauga and domed stadium in Toronto	Lockport (Eramosa)	Dolostone	—	—	120.0	63.0	—	0.0 h
		Lockport (Goat Island)	Dolostone	—	—	200.0	62.0	—	0.0 h
		Lockport (Gasport)	Shaly Limestone	—	—	120.0	27.0	—	0.08 h
		De Cew	Dolostone/Mudstone	—	—	74.0	57.0	—	0.04 h
		Rochester	Shale	—	—	85.0	23.0	—	0.07 h
		Irondequoit	Limestone	—	—	90.0	60.0	—	—
		Reynolds	Dolostone	—	—	106.0	40.0	—	—
		Grimsby	Sandstone	—	—	132.0	42.0	—	—
			Shale	—	—	25.0	8.0	—	0.27 h

	Formation	Rock type							[3]
Southern Ontario — Thorold tunnel, wheel pits in the Canadian Niagara falls and Toronto power g.s., heart lake tunnel in Mississauga, intake tunnel of Darlington g.s., Scotia plaza in Mississauga and domed stadium in Toronto	Power Glen	Sandstone	–	–	158.0	52.0	–	–
		Shale	–	–	26.0	9.0	–	0.17 h
	Whirlpool	Sandstone	–	–	190.0	55.0	–	–
	Queenston	Shale	–	–	30.0	10.0	–	0.30 h
	Georgian Bay	Shale	–	–	20.0	4.0	–	0.15 h
	Blue Mountain	Shale	–	–	27.0	2.0	–	0.15 h
	Collingwood	Black shale	–	–	80.0	20.0	–	0.00 h
		Grey mud-stone	–	–	58.0	10.0	–	0.15 h
	Lindsay	Shaly lime-stone	–	–	110.0	46.0	–	0.05 h
	Verulam	Limestone (Shaly interbeds)	–	–	23.0	57.0	–	0.05 h
	Bobcay-geon	Shaly lime-stone	–	–	78.0	56.0	–	–
	Gull River	Limestone	–	–	143.0	63.0	–	0.00 h

									[3]
Southern Ontario	Thorold tunnel, wheel pits in the Canadian Niagara falls and Toronto power g.s., heart lake tunnel in Mississauga, intake tunnel of Darlington g.s., Scotia plaza in Mississauga and domed stadium in Toronto	Shadow Lake	Sandstone	—	—	60.0	21.0	—	—
		Pre Cambrian	Medium grained	—	—	190.0	60.0	—	0.00 h
		Granitic	Coarse grained	—	—	140.0	46.0	—	0.00 h
		Gneiss	Gneiss bands	—	—	90.0	46.0	—	—

Region	Location	Formation	Rock type							Ref.
				—	—	—	—	—	—	
South-ern On-tario	Mississauga, Pickering, Bowman-ville, Wesleyville and Port Hope In Ontario	Cobourg	Argil-laceous Limestone	—	0.04 - 2.0 d 3.0 -10.0 b	22.0 - 140.0	10.0 - 67.0	0.1 - 0.6	—	[20]
			Colling-wood shale	—	—	27.0 - 132.0	2.0 - 31.0	0.2 - 0.3	—	
		Sherman Fall	Shale	—	0.1 - 3.0 d 1.0 - 12.0 b	23.0 - 69.0	1.0 - 73.0	0.1 - 0.4	—	
			Interbedded limestone	—	0.1 - 3.0 d 1.0 - 12.0 b	71.0 - 161.0	1.0 - 73.0	0.1 - 0.4	—	
		Kirkfield and Co-boconk	—	—	—	34.0 - 115.0	13.0 - 64.0	—	—	
Quebec/ Beauchene	Tunnel In Lake Beauchene, Quebec	—	Gneiss W. Mica/ Quartz	64.0	—	—	34.5	—	—	[13] [34]
Quebec/ Church-hill falls	Cavern adit in Church-hill falls	—	Gneissic	305.0	—	—	48.0	—	—	[35]
Mani-toba/ Pinawa	Under-ground research laboratory (URL)	—	Granite	336.6 - 515.0	—	167.0	—	—	—	[9] [26]
Mani-toba/ Pinawa		—	Granite	470.1 - 851.3	—	—	15.6 - 25.8	—	—	[38]
South-ern Illinois	Oil field-deep boring	—	Carbonate	99.1	—	—	14.0	—	—	[17]

Location		Site		Rock type					Reference
Minnesota/ St. Cloud	—	Quarry	—	Granite	—	—	47.0	—	[45]
New York/ Alma Township	—	Oil field-deep boring	—	Sandstone	502.9	—	7.0	—	[19]
New York/ Briarcliff Manor	—	Outcrop	—	Gneiss	5.6 - 13.1	—	3.0 - 52.0	—	[13]
New York/ Nyack	—	Outcrop	—	Diabase	0.2 - 0.5	—	19.6	—	[41]
New York/ Niagara gorge	—	Outcrop	—	Dolomite	0.2 - 6.7	—	24.0	—	[13] [40]
New York/ Rochester	—	Sewer system	—	Dolomite	—	—	50.7 - 91.7	—	[42]
New York/ Somerset	—	Outcrop	—	Sandstone	8.5	—	17.0	—	[13] [43] [44]

Location	Type	Formation	Rock type							Reference
New York/Sterling	Outcrop	—	Sandstone	10.1 - 32.3	—	—	33.0	—	—	[13][43][44]
Ohio	Boring	—	Shale	10.3 - 18.6	—	13.0 - 28.0	—	—	—	[13]
Ohio/Barberton	Mine	—	Limestone	701.0	—	55.0 - 67.0	—	—	—	[21]
Ohio/Bellefountaine	Quarry	Gasport	Limestone/Dolomite	0.2 - 1.0	—	34.8	—	—	—	[13]
Ohio/Falls Township	Oil Field - Deep boring	—	Sandstone	808.0	—	10.0	—	—	—	[17]
Ohio/Hocking State Forest	Outcrop	—	Sandstone	0.9 - 1.2	—	7.8	—	—	—	[14]
Ohio/Kenton	Quarry	—	Limestone/Dolomite	0.2 - 1.0	—	34.8	—	—	—	[13]
Ohio/Lima	Quarry	—	Limestone/Dolomite	0.2 - 1.0	—	34.8	—	—	—	[13]
Ohio/Sydney	Quarry	—	Limestone/Dolomite	0.2 - 1.0	—	34.8	—	—	—	[13]

Wis-consin/ Mon-tello	Deep Bor-ing	–	Granite	75.0 - 188.1	–	–	–	52.0 - 56.0	–	[46]

d: result from direct tension test; b: results from Brazilian test; v: results from vertically cored samples/or measurements in the vertical direction; h: results from horizontally cored samples/or measurements in the horizontal direction ; i: results from inclined 45° cored samples with respect to the bedding planes.

The tensile strength of intact rock is measured in a laboratory either directly with the direct tension test or indirectly with the indirect tension test, which is commonly known as a Brazilian test or a split test. In the direct tension test, a cylindrical rock specimen is subjected to a direct uniaxial tensile stress along its longitudinal axis until failure. In the Brazilian test, the indirect tensile strength of the rock is measured on disc specimens by applying a compressive stress across the disc perimeter until failure. The failure occurs along the diameter of the disc specimen in a biaxial state of stress where one principal stress is highly compressive. In general, the indirect tensile strength of rock measured from the Brazilian test is higher than the tensile strength of the same rock measured from the direct tension test.

The compressive strength, elastic (Young's) modulus, and Poisson's ratio of intact rocks are all measured in a laboratory either through a uniaxial compression test or a triaxial compression test. In the uniaxial compression test, a cylindrical rock specimen is subjected to a compressive stress along its longitudinal axis until failure occurs, while in the triaxial compression test, failure is similarly induced when the cylindrical rock specimen is subjected to a specific value of confining pressure. In both tests, electronic strain gauges are affixed onto the specimen, parallel and perpendicular to the longitudinal axis of the specimen, to measure the axial and diametric deformations during the tests. The elastic theory relationships are then used to calculate the elastic modulus and Poisson's ratio.

The strength and stiffness characteristics of intact rock specimens extracted from different rock formations in Southern Ontario and the neighbouring regions were extensively investigated over the past decades [4] [18] [24] -[26]. However, the in-situ medium (i.e. the rock mass) comprises of intact rock blocks that are separated by discontinuities such as joints, fissures and faults [27]. These discontinuities have a great influence on the overall strength characteristics of the rock mass, and therefore they have to be prudently considered in evaluating the overall strength of the rock mass. The rock mass modulus can be measured in-situ by recording the deformation in the diameter of a pre-drilled monitoring hole through the rock mass while extending the tunnel excavations. The deformation is recorded using an extensometer probe that is affixed at the bottom of the monitoring hole. Another field test method was developed in 1987 by Lo, Yung and Lukajic [25] to measure the rock mass modulus at the surface of the excavated rock. In principle, the developed method consisted of measuring the variation in the diametric distance between each opposite pair of pre-glued props into pre-drilled holes from the surface of the rock layer, in a rosette pattern, while extending a central hole into the rock layer from the surface. The elastic theory was then used to calculate the rock mass modulus [25]. The developed method was used to measure the rock mass modulus of the limestone layer at the intake and discharge tunnels of Darlington Generating Station, east of Toronto. The values of the measure rock modulus from this method were consistent with those evaluated from extensometer measurements in the tunnels.

As mentioned before, the strength data presented in Table 2 were assembled from laboratory tests performed on intact rock specimens of samples extracted from variable depths and diversity of rock formations in the concerned area. In general, the dolomitic limestone of Lockport formation possesses the highest uniaxial compression strength of 199 - 246 MPa among all other rocks in Southern Ontario [5]. The sandstone of Whirlpool formation and the dolostone of Lockport formation exhibit uniaxial compression strength of 190 MPa and 200 MPa, respectively [3]. The black shale of Collingwood formation and the Rochester shale exhibit a high uniaxial compression strength of 80 - 85 MPa in contrast to other shales in Southern Ontario, such as Georgian Bay, Grimsby, Power Glen, Blue Mountain, and Queenston in which the uniaxial compression strength ranges between 20 - 30 MPa [3]. Moreover, most of the sedimentary rocks of Southern Ontario possess anisotropy in their uniaxial compression strength, with respect to the bedding planes.

As can be seen from Table 2, the available data of the tensile strength of rocks in the specified area are very limited. However, it is reported that the tensile strength of Queenston shale from different sites in Southern Ontario varies between 1MPa to 15 MPa in contrast to Sherman Fall shale where the tensile strength is 0.1 - 3 MPa [20]. It is reported that the dolostone and mudstone of De Cew formation possess a tensile strength of 5 MPa [20].

The elastic modulus of siderite and tuff in Wawa, Ontario was reported as 67.6 - 118.0 GPa and 68.3 - 115.8 GPa [28], respectively. The quartzite and sandstone of Elliot Lake, Ontario possess an elastic modulus of 80.0 GPa and 76.0 GPa respectively [18] [29], while the shales, in general, possess an average elastic modulus of 10.0 GPa [2] -[4] [7] [8] [20]. On the other hand, the Poisson's ratio of rocks in Southern Ontario was ranging from 0.13 for Georgian Bay shale [5] to 0.6 for argillaceous limestone of Cobourg formation [20]. Moreover, most of the sedimentary rocks of Southern Ontario possess anisotropy in their strength and stiffness properties, with respect to the bedding planes.

As stated earlier, the presented data in Table 2 are based on laboratory tests that were performed on freshly recovered intact rocks from the ground. In practice, the rocks at the surfaces of the underground tunnel excavations are actually exposed either to water or other drilling fluids, such as bentonite slurry or synthetic polymers solutions as part of the construction process for the buried infrastructures. These drilling fluids are used as lubricant to facilitate the drilling process through the rock mass or to convey the excavated rocks. As mentioned before, there is lack of information with regard to the influence of the exposure of rocks to the drilling fluids near the surfaces of excavation on the strength characteristics of these rocks, therefore, the influence of these drilling fluids on the strength and stiffness characteristics of rocks in Southern Ontario is under ongoing investigation at Western University.

Intact Rock Time-Dependent Deformation Properties

The swelling potential of rocks is an important factor in designing underground structures and has a significant influence on the stability of these structures. As proposed by Lo, Palmer and Quigly [7], the swelling potential in the swelling rocks can be defined as the swelling strain per log cycle of time and it can be calculated through the free swell test. In the free swell test, the intact rock specimen is submerged in water and allowed to expand freely in all directions while the swelling strain is measured in three orthogonal directions [7]. The horizontal swell strain is measured in the direction parallel to the bedding planes of the rock sample, while the vertical swell strain is measured in the direction perpendicular to the bedding planes. The swelling potential values measured in the vertical and horizontal directions with respect to the bedding planes of different rock formations in Southern Ontario and the neighbouring regions are presented in Table 2.

As can be seen in Table 2, most of the shaly rock formations exhibit anisotropy in their swelling behaviour in the direction parallel and perpendicular to the bedding planes [6] -[8]. For example, the Queenston shale from Niagara Falls exhibits swelling potential of 0.37% - 0.54% in the vertical direction and 0.22% - 0.34% in the horizontal direction [6]. The Georgian Bay shale from different sites in Southern Ontario indicates swelling potential of 0.2% - 0.22% in the vertical direction and 0.03% - 0.14% in the horizontal direction [7]. The Rochester shale exhibits relatively small swelling potential averaging 0.16% and 0.07% in the vertical and horizontal directions, respectively [7]. In general, the limestone displays zero swelling potential due to their high calcite content, however, some shaly limestone such as Gasport shaly limestone exhibits swelling potential of 0.08% in both horizontal and vertical directions [7].

Lee and Lo [8] investigated the swelling mechanism of shales in Southern Ontario by submerging the shale specimens in water with varying salt concentrations. Based on the results of their investigations, they suggested that the swelling mechanism of shales in this region was based on the process of osmosis and diffusion which occurred between the rock pore water and the ambient fluid. It was concluded that swelling occurs if three conditions are met: i) relief of initial stress, ii) accessibility of water and iii) an outward salt concentration gradient from pore fluid exists. They assumed that swelling may or may not occur if only one or two of these conditions are met. Although the swelling behaviour of shales in Southern Ontario was extensively investigated in water [3] [6] -[8], there is lack of information with respect to swelling behaviour of these shales in drilling fluids, such as bentonite slurry and synthetic polymers solutions.

Dynamic Properties of Rocks

The compressional wave velocity, shear wave velocity, dynamic Poisson's ratio and dynamic modulus of different rock formations in Southern Ontario were compiled and presented in Table 3. The compressional wave and the shear wave velocities were measured on intact rock specimens and the dynamic Poisson's ratio and the dynamic modulus were calculated using the fundamental equations for torsional vibration [5] [6] [25].

In general, the presented data revealed anisotropy in the dynamic behaviour of the sedimentary rocks in Southern Ontario. For the same rock formation, the value of the dynamic modulus in the direction parallel to the bedding planes is higher than that in the direction perpendicular to the bedding planes. It should be noted that the presented dynamic properties are obtained for intact rock specimens. However, the effects of saturation in drilling fluids such as bentonite slurry and synthetic polymers solution on the dynamic properties of rocks still need to be investigated.

Table 3. Dynamic properties of intact rocks.

Province/State/City	Project	Rock Formation	Rock Type	Depth/Elevation (m)	Mass Density (Mg/m³)	Compressive Wave Velocity (Km/s)	Shear Wave Velocity (Km/s)	Dynamic Poisson's Ratio vdy.	Dynamic Modulus Edy. (GPa)	Source of Data
Southern Ontario	Research Program For The National Research Council of Canada, Different Sites in Southern Ontario	Lockport (Gasport)	Shaly limestone	159.94	2.68 - 2.69 v	–	–	–	44.3 - 67.5 v	[5]
		Lockport (Gasport)	Shaly limestone	162.05	2.68 - 2.76 h	–	–	–	63.3 - 71.0 h	
		Lockport (Gasport)	Fossiliferous limestone	159.23	2.71 v	–	–	–	66.8 v	
		Lockport (Gasport)	Fossiliferous limestone	157.33	2.72 h	–	–	–	73.1 h	
		Georgian Bay	Shale	15.33	2.55 v	–	–	–	19.2 v	
		Georgian Bay	Shale	10.17	2.60 h	–	–	–	38.2 h	
		Georgian Bay	Shale	12.1	2.54 i	–	–	–	19.0 i	
		Georgian Bay	Shale	22.76	2.53 v	–	–	–	27.4 v	
		Collingwood	Black shale	18.49	2.53 - 2.56 h	–	–	–	51.3 - 58.4 h	
		Collingwood	Black shale	16.99	2.58 i	–	–	–	37.3 i	
		Collingwood	Grey shale	23.34	2.6 v	–	–	–	4.9 v	
		Collingwood	Grey shale	23.27	2.61 - 2.64 h	–	–	–	42.2 - 49.2 h	

Region	Source	Location	Rock type			[5]					[5]
Southern Ontario	Research Program For The National Research Council of Canada, Different Sites in Southern Ontario	Collingwood	Grey shale	23.29	2.6 i	—	—	—	—	—	—
		Collingwood	Shaly limestone	12.93 - 26.06	2.68	—	—	—	—	—	—
		Trenton	Limestone	35.41	2.68 v	—	—	—	—	—	—
		Trenton	Limestone	35.53	2.68 h	—	—	—	—	—	—
		Trenton	Limestone	35.48	2.85 i	—	—	—	—	—	—
		Rochester	Shale	26.37	2.77 v	—	—	—	—	—	38.7 v
		Rochester	Shale	26.24 - 26.5	2.68 - 2.72 h	—	—	—	—	—	39.4 h
		Rochester	Shale	26.29	2.74 i	—	—	—	—	—	21.8 i
		Lockport (Goat Island)	Dolomite	169.37	2.76	—	—	—	—	—	61.9 v
		Lockport (Goat Island)	Dolomite	168.8	2.76	—	—	—	—	—	70.3 - 80.2 h
		Lockport (Goat Island)	Dolomite	169.21	2.77	—	—	—	—	—	74.5 i
		Lockport (Gasport)	Dolomitic limestone	165.66	2.72 v	—	—	—	—	—	73.8 v
Southern Ontario	Research Program For The National Research Council of Canada, Different Sites in Southern Ontario	Lockport (Gasport)	Dolomitic limestone	165.57	2.76 h	—	—	—	—	—	70.3 - 86.5 h
		Lockport (Gasport)	Dolomitic limestone	165.74	2.72 i	—	—	—	—	—	69.6 i
		Lockport (Gasport)	Dolomitic limestone/ Limestone	164.06	2.72 v	—	—	—	—	—	47.8 v
		Lockport (Gasport)	Dolomitic limestone/ Limestone	164.17	2.71 - 2.72 h	—	—	—	—	—	53.4 - 66.1 h
		Lockport (Gasport)	Dolomitic limestone/ Limestone	164..01	2.76 i	—	—	—	—	—	60.5 i

Ontario/ Toronto	Darlington intake tunnel	Whitby	Shaly limestone	83.4	2.58 - 2.70	5.1 - 5.12 v	1.01 - 2.49	0.34 - 0.37 v	39.6 - 43.6 v	[4]
				84.4 - 84.7	—	4.92 - 5.13 h	—	0.25 - 0.27 h	52.9 - 54.6 h	[25]
Ontario / Niagara Falls	Sir Adam Beck Niagara generating station (SABNGS) No. 3	Queenston	Shale	95.64 - 114.33	2.66 - 2.68	3.48 - 4.28	—	—	—	[6]

v: results from vertically cored samples/or measurements in the vertical direction; h: results from horizontally cored samples/ or measurements in the horizontal direction; i: results from inclined 45° cored samples with respect to the bedding planes.

SUMMARY AND CONCLUSIONS

A comprehensive review of the available literature on the geo-mechanical properties of rock formations in Southern Ontario and the neighbouring regions (New York, Pennsylvania, Ohio, Michigan, Indiana, Illinois, Wisconsin, and Minnesota) was performed. The available data on the measured in-situ stresses and the direction of major principal stress, strength and stiffness properties, time-dependent deformation properties, and dynamic properties of different rocks from that literature were compiled. The presented data can serve as a preliminary source of information for any prospective study of the geo-mechanical properties of the rocks in Southern Ontario and the neighbouring regions.

From this compiled data, the following conclusions can be drawn:

- The value of the initial in-situ horizontal stress in rock formations of Southern Ontario and the neighbouring regions varies from a relatively small amount, <1 MPa, to a considerably high amount, >100 MPa, depending on the rock formation, depth and inter-bedded layers in the rock mass. For depths up to 30 m where most of the engineering projects are located, the in-situ horizontal stresses in rocks of Southern Ontario and the neighbouring regions are ranging between −4.87 MPa to 38.13 MPa, while for depths greater than 30 m and up to 1000 m where the mining projects are located, the in-situ horizontal stresses are ranging between 1.59 MPa to 85.7 MPa. Moreover, the in-situ horizontal stresses are considerably high for depths greater than 1000 m where the hydrocarbons projects are located, ranging from 42.0 MPa to as high as 135.0 MPa.

- Among shales of Southern Ontario and the neighbouring regions, the Queenston shale of Niagara Falls region exhibits highest swelling potential of 0.37% - 0.54% in the vertical direction and 0.22% - 0.34% in the horizontal direction, with respect to the bedding planes.

- The sedimentary rocks and shales in particular, possess considerable anisotropy in their strength, time-de- pendent deformation and dynamic properties, relative to the bedding planes.

- Although the swelling behaviour of rocks in Southern Ontario and the neighboring regions was extensively investigated using water as an ambient solution, there is a lack of information with respect to the time-depen- dent deformation behaviour of these rocks in fluids such as bentonite slurry and synthetic polymers solution. For most of the tunnel drilling process through the rock mass, other than blasting, fluids such as bentonite slurry and synthetic polymers solutions are used either to convey the excavated materials or to lubricate the annulus of the excavated tunnel. Therefore, it is quite indispensable to investigate the influence of these fluids on the strength, time-dependent deformation and dynamic characteristics of these rocks, which is the topic of the ongoing research at Western University.

ACKNOWLEDGEMENTS

This Research is being performed under the umbrella of the Geotechnical Research Centre (GRC) in Western University. The authors would like to thank Ward & Bark Microtunnelling Ltd. For their financial support, the authors also would like to thank Dr. K. Y. Lo for his valuable guidance.

REFERENCES

1. Lo, K.Y. (1978) Regional Distribution of in Situ Horizontal Stresses in Rocks of Southern Ontario. Canadian Geo- technical Journal, 15, 371-381. http://dx.doi.org/10.1139/t78-034

2. Lo, K. Y., and Yuen, C.M.K. (1981) Design of Tunnel Lining in Rock for Long Term Time Effects. Canadia Geotech- nical Journal, 18, 24-39. http://dx.doi.org/10.1139/t81-004

3. Lo, K.Y. (1089) Recent Advances in Design and Evaluation of Performance of Underground Structures in Rocks. Tunnelling and Underground Space Technology, 4, 171-183.http://dx.doi.org/10.1016/0886-7798(89)90050-3

4. Lo, K.Y., Cooke, B.H. and Dunbar, D.D. (1987) Design of Buried Structures in Squeezing Rock in Toronto, Canada. Canadian Geotechnical Journal, 24, 232-241.http://dx.doi.org/10.1139/t87-028

5. Lo, K.Y. and Hori, M. (1979) Deformation and Strength Properties of Some Rocks in Southern Ontario. Canadian Geotechnical Journal, 16, 108-120.http://dx.doi.org/10.1139/t79-010

6. Lo, K.Y. and Lee, Y.N. (1990) Time-Dependent Deformation Behaviour of Queenston Shales. Canadian Geotechnical Journal, 27, 461-471. http://dx.doi.org/10.1139/t90-061

7. Lo, K.Y., Palmer, J. H. L. and Quigly, R.M. (1978) Time-Dependent Deformation of Shaly Rocks in Southern Ontario. Canadian Geotechnical Journal, 15, 537-547.http://dx.doi.org/10.1139/t78-057

8. Lee, Y.N. and Lo, K.Y. (1993) The Swelling Mechanism of Queenston Shale. Canadian Tunnelling, 75-97.

9. Hefney, A., Lo, K.Y. and Huang, J.A. (1996) Modelling of Long-Term

Time-Dependent Deformation and Stress De- pendency of Queenston Shales. Tunnelling Association of Canada Annual Publication, 115-146.

10. Map of Southern Ontario and Neighbouring Regions, Google Maps. https://maps.google.ca

11. Terzaghi, K. (1962) Measurement of Stresses in Rock. Géotechnique, 12, 105-124.http://www.icevirtuallibrary.com/content/article/10.1680/geot.1962.12.2.105http://dx.doi.org/10.1680/geot.1962.12.2.105

12. Von Schonfeldt, H. and Fairhurst, C. (1972) Field Experiments on Hydraulic Fracturing. American Institute of Mining Engineers, 253, 69-77.

13. Lindner, E.N. and Halpern, J.A. (1978) In-Situ Stress in North America: A Compilation. International Journal of Rock Mechanics and Mining Sciences and Geomechanics Abstracts, 15, 183-203. http://dx.doi.org/10.1016/0148-9062(78)91225-1

14. Rough, R.L. and Lambert, W.G. (1971) In-Situ Strain Orientations: A Comparison of Three Measuring Techniques. Report of Investigation, 7575, US Department of Interior, Bureau of Mines, USA. http://books.google.ca/books?id=DgXoGQAACAAJ

15. Haimson, B.C. (1976) The Hydrofracturing Stress Measuring Technique-Method and Recent Field Results in the US. The Proceedings of the International Society of Rock Mechanics Symposium, Sydney, 23-30.

16. Lo, K.Y. and Hefney, A. (1993) The Evaluation of In-Situ Stresses by Hydraulic Fracturing Tests in Anisotropic Rocks with Mixed-Mode Fractures. Canadian Tunnelling, 59-73.

17. Haimson, B. and Fairhurst, C. (1970) In-Situ Stress Determination at Great Depth by Means of Hydraulic Fracturing. Rock Mechanics— Theory and Practice. The Proceedings of the 11th Symposium on Rock Mechanics, 16-19 June 1969, Berkeley, 559-584.

18. van Heerden, W.L. and Grant, F. (1967) A Comparison of Two Methods for Measuring Stress in Rock. International Journal of Rock Mechanics and Mining Sciences & Geomechanics Abstracts, 4, 367-382. http://dx.doi.org/10.1016/0148-9062(67)90028-9

19. Haimson, B. and Stahl, E.J. (1969) Hydraulic Fracturing and the Extraction of Minerals through Wells. The Proceedings of the 3rd Symposium on Salt, Northern Ohio Geological Society, Cleveland, 421-432.

20. Lam, T., Engelder, T., Leech, R.E.J. and Jensen, M. (2011) Regional Geomechanics–Southern Ontario. Nuclear Waste Management Organization and AECOM Canada Ltd., Technical Report No. NWMO

DGR-TR-2011-13.

21. Obert, L.A. (1962) In-situ Determination of Stress in Rock. Mining Engineering, 14, 51-58.

22. Buchbinder, G.G.R., Nyland, E. and Blanchard, J.E. (1965) Measurement of Stress in Bore-Holes, in Drilling for Scientific Purposes. Proceedings of the International Upper Mantle Symposium, Ottawa, 2-3 September 1965, 85-93.

23. Fairhurst, C. (2004) Nuclear Waste Disposal and Rock Mechanics: Contributions of the Underground Research Laboratory (URL), Pinawa, Manitoba, Canada. International Journal of Rock Mechanics and Mining Sciences, 41, 1221-1227.http://dx.doi.org/10.1016/j.ijrmms.2004.09.001

24. Palmer, H.L. and Lo, K.Y. (1976) In Situ Measurements in Some near Surface Rock Formation—Thorold, Ontario. Canadian Geotechnical Journal, 13, 1-7.http://dx.doi.org/10.1139/t76-001

25. Lo, K.Y., Yung, C.B. and Lukajic, B. (1987) A Field Method for the Determination of Rock-Mass Modulus. Canadian Geotechnical Journal, 24, 406-413.http://dx.doi.org/10.1139/t87-051

26. Hefney, A. and Lo, K.Y. (1995) Interpretation of Initial Stresses from Hydraulic Fracturing Tests at AECL's Underground Research Laboratory, Manitoba. Tunnelling Association of Canada Annual Publication, 123-134.

27. Hoek, E. (2001) Big Tunnels in Bad Rocks—2000 Terzaghi Lecture. American Society of Civil Engineers Journal of Geotechnical and Geoenvironmental Engineering, 127, 726-740.http://www.rocscience.com/hoek/references/H2001b.PDF

28. Herget, G. (1973) Variation of Rock Stresses with Depth at a Canadian Iron Mine. International Journal of Rock Mechanics and Mining Sciences & Geomechanics Abstracts, 10, 37-51. http://dx.doi.org/10.1016/0148-9062(73)90058-2

29. Eisbacher, G.H. and Bielenstein, H.U. (1971) Elastic Strain Recovery in Proterozonic Rocks Near Elliot Lake, Ontario. Journal of Geophysical Research, 76, 2012-2021.http://dx.doi.org/10.1029/JB076i008p02012

30. Canada. Mines Branch, Coates, D.F., Grant, F. and Heerden, W.L. (1968) Stress Measurements at Elliot Lake. Canada Mines Branch Reprint Series, R. Duhamel, 603-613.http://books.google.ca/books?id=13N_tgAACAAJ

31. Franklin, J.A. and Hungr, O. (1978) Rock Stresses in Canada: Their Relevance to Engineering Projects. Rock Mechanics, 6, 25-46.

32. Lo, K.Y., Lukajic, B., Yuen, C.M.K. and Hori, M. (1982) In-Situ Stresses in a Rock Overhang at the Ontario Power Generating Station, Niagara Falls. Proceedings of the 23rd Symposium on Rock Mechanics, Berkeley, 25-27 August 1982, 343-352.

33. Moruzi, G.A. (1968) Application of Rock Mechanics in Mine Planning and Ground Control. Current Rock Mechanics, Chap. 3, Studies at Falconbridge. Canadian Mining Journal, 89, 12-15.

34. Metaltech Inspection Ltd. (1970) Rapport D'Investigation, Measure Des Constraintes En Place. Contract No. 1009-61, Commission Hydro-Electrique du Quebec.

35. Benson, R.P., Kierans, T.W. and Sigvaldason, O.T. (1970) In-Situ and Induced Stresses at the Churchill Falls Underground Powerhouse, Labrador. Proceedings of the 2nd Congress of the International Society for Rock Mechanics, 4, 821-832.

36. Hefney, A. and Lo, K.Y. (1992) The Interpretation of Horizontal and Mixed-Mode Fractures in Hydraulic Fracturing Test in Rocks. Canadian Geotechnical Journal, 29, 902-917.http://dx.doi.org/10.1139/t92-102

37. Fairhurst, C. (2003) Stress Estimation in Rock: A Brief History and Review. International Journal of Rock Mechanics and Mining Sciences, 40, 957-973.http://dx.doi.org/10.1016/j.ijrmms.2003.07.002

38. Thompson, P.M. and Chandler, N.A. (2004) In-Situ Rock Stress Determinations in Deep Boreholes at the Underground Research Laboratory. International Journal of Rock Mechanics and Mining Sciences, 41, 1305-1316.http://dx.doi.org/10.1016/j.ijrmms.2004.09.003

39. Sbar, M.L. and Sykes, L.R. (1973) Contemporary Compressive Stress and Seismicity in Eastern North America: An Example of Intra-Plate Tectonics. Geological Society of America Bulletin, 84, 1861-1882. http://dx.doi.org/10.1130/0016-7606(1973)84<1861:CCSASI>2.0.CO;2

40. Sellers, J.B. (1969) Strain Relief Overcoring to Measure in-Situ Stresses. Niagara Falls Project, Corps of Engineers, Buffalo District.

41. Hooker, V.E. and Johnson, C.F. (1977) In-Situ Stresses along the Appalachian Piedmont. Proceedings of the 4th Canadian Symposium on Rock Mechanics, Ottawa, 29-30 March 1967, 137-155.

42. Goldberg, Z. and Dunnicliff and Associates (1976) Report on in-Situ Stress Measurements, Genessee River Interceptor, Southwest Rochester, NewYork, File No. 1661, Firelands Sewer and Water Construction Company, New York.

43. Dames and Moore Consultants (1973) In-Situ Stress Measurement.

Report of Investigation, North Anna Power Project.

44. Dames and Moore Consultants (1974) Geological Report. Report No. 4852-002-18, Limerick Generating Station.

45. Bonnechere, F. (1969) A Comparative Field Study of Rock Stress Determination Techniques. Missouri River Division, Report No. 68101, Corps of Engineers, Omaha.

46. Foundation Sciences Incorporation (1971) Bear Swamp Project-Rock Mechanics Studies. New England Power Service Company, Westborough.

47. Lo, K.Y. and Lukajic, B. (1984) Predicted and Measured Stresses and Displacements around the Darlington Intake Tunnel. Canadian Geotechnical Journal, 21, 147-165.http://dx.doi.org/10.1139/t84-012

Chapter 2

REMOTE SENSING ROCK MECHANICS AND EARTHQUAKE THERMAL INFRARED ANOMALIES

Lixin Wu[1,2] and Shanjun Liu[2]

[1]Academy of Disaster Reduction & Emergency Management, Beijing Normal University, Beijing,, China
[2]Institute for Geo-informatics & Digital Mine Research, Northeastern University, Shenyang, China

INTRODUCTION

Rock fracturing is the cause of many geo-hazards including tectonic earthquake (EQ), rock burst, rock sloping and rock pillar failure. Radiation signals such as acoustic emission, radio frequency emission and electromagnetic (EM) radiation from loaded deforming rock, are able to provide useful information for monitoring, interpreting and predicting rock fracturing (Renata, 1977, Brady and Rowell, 1986, Yamada et al., 1989, Martelli et al., 1989). Based on thermo-elastic theory, thermo-elastic stress analysis (TSA) and stress pattern analysis by thermal emission (SPATE) were developed for the stress measurement of solid materials, including homogeneous metal, macromolecular and composite materials, respectively in 1960's and 1970's (Mounatin and Webber, 1978). Luong applied thermovision to study experimentally the damage processes of concrete and rock (Luong, 1990), but no reach to the remote sensing on geo-hazards.

In the experiments for investigating the mechanism of satellite thermal infrared (TIR) anomaly before tectonic EQ (Gorny et al., 1988, Qiang et al., 1990), it was discovered that there do exist TIR anomaly before rock fracturing (Geng et al., 1992). Later, it was furthermore discovered that there are obvious TIR features as precursors of rock fracturing, and that the loaded stress around 0.79 σ_c can be taken as a precaution index for the stability monitoring of loaded rocks (Wu and Wang, 1998). To explore the laws of infrared radiation (IRR) variation in the process of rock loading, deforming and fracturing, and to reveal the possible mechanism of satellite TIR anomaly before EQ, a large amount of IRR imaging experiments on rock loaded to fracturing were conducted in China (Wu et al., 2000, 2001, 2002, 2003,2004a, 2004b, 2004c, 2004d, 2006a, 2006b; Deng et al., 2001, Liu et al., 2002). Hence, a new intersection discipline, Remote Sensing Rock Mechanics (RSRM), which takes Remote Sensing, Rock Mechanics, Rock Physics and Informatics as its foundations and serves for remote sensing on geo-hazards, was originated (Geng et al., 1992; Wu et al., 2000).

Based on retrospection to past experiments on RSRM, it was pointed out that there are two IRR anomalies, being IRR image anomaly and IRR temperature curve anomaly respectively, can act as rock fracturing precursors. The average IRR temperature (AIRT), being the integral reflection of surface IRR energy, is applied as a quantitative index to study the temporal evolution of IRR from loaded rock and to seek for the potential precursors of rock fracturing. The temporal evolution of AIRT are the comprehensive effect of a series of physical-mechanical processes inside a loaded rock, such as rock thermo-elastic acting, pore gas desorbing & escaping, fractures producing & extending, rock frictionating, heat transferring and environment radiation. The thermo-elastic effect and the frictional thermal are two of the main mechanisms of increased IRR from loaded rock. RSRM experiments had revealed the laws of changed IRR from loaded rock and provided scientific interpretations for the mechanisms of satellite TIR anomaly before tectonic EQs of Ms>5.5.

REMOTE SENSING ROCK MECHANICS EXPERIMENTS

Experiment Methods and Tools

The typical RSRM experiment is comprised of a loader (uni-axial or bi-axial), an infrared imager and rock samples. As in Figure 1, a bi-axial loader was applied for loading along two directions, and an infrared imager was applied to detect the surface IRR from loaded rock. The maximum imaging rate of the imager is 60f/s, and the recording rate was usually set as 1f/s to record the IRR images continuously. Usually, tectonic EQ might be resulted from the suddenly fracturing of compressively-sheared crust rock, the suddenly breaking of faults at disjointed zones, the suddenly sliding of compressively-sheared faults or the stability losing of compressively loaded intersected faults. To simulate the different mechanisms of rock fracturing and EQ, several typical loading schemes were applied as in Figure 1.

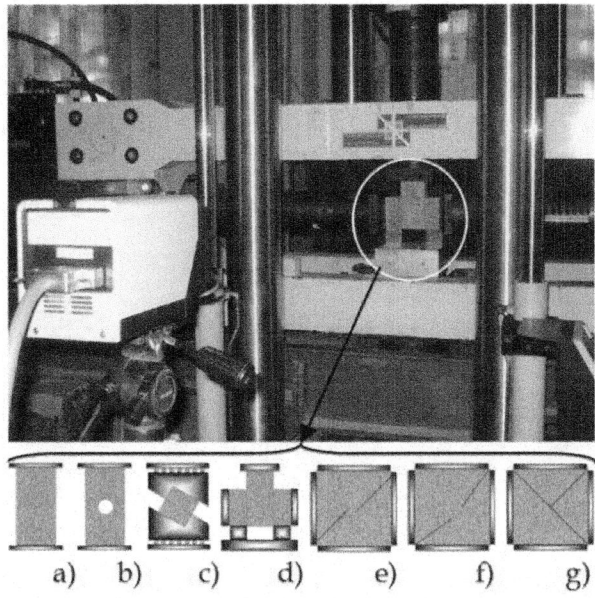

Figure 1. RSRM experiment schemes to simulate different mechanisms of rock fracturing or tectonic EQ: a) uni-axially load on a standard cylinder rock sample; b) uni-axially load on a cylinder rock sample with a central hole; c) compressively-sheared load on a hexahedral rock sample; d) bi-axially load on three jointed rock samples to frictional sliding; e) bi-axially load on a damage rock sample with en echelon faults; f) bi-axially load on a damage rock sample with disjointed faults; and g) bi-axially load on three jointed rock samples simulating intersected faults.

Rock Fracturing Precursor: IRR Image Anomaly

Uni-axially Loaded Rock

Lots of rock samples made from coal, ironstone, sandstone, marble, limestone, granite, granodiorite, gabbro and gneiss were uni-axially loaded and thermal imaging detected. The sample size was standard of diameter and length, respectively, 50 and 100mm. It was discovered that the IRR images of the uni-axially loaded rock have different features for different fracturing pattern (Wu et al., 2006a). As in Figures 2~4, there are three fracturing patterns, "X"-shaped, "//"-shaped and "|"-shaped respectively, occurred in our experiments. The "X"-shaped and "//"-shaped positive IRR abnormal strips foretell the coming of "X"-shaped shearing fracturing and the coming of "//"-shaped shearing fracturing respectively, while the "|"-shaped negative IRR abnormal strip foretells the coming of tensile fracturing.

The "X"-shaped positive IRR abnormal strips generated with loading along the "X"-shaped shearing zone before peak stress, and got distinguished after peak stress, as in Figure 2. The rock sample got finally fractured along the "X"-shape shearing zone. The evolution of IRR abnormal strip had also reflected the fracturing being not symmetrical upper-and lower, in that the upper part was clear with higher temperature, while the lower part is fuzzy with lower temperature.

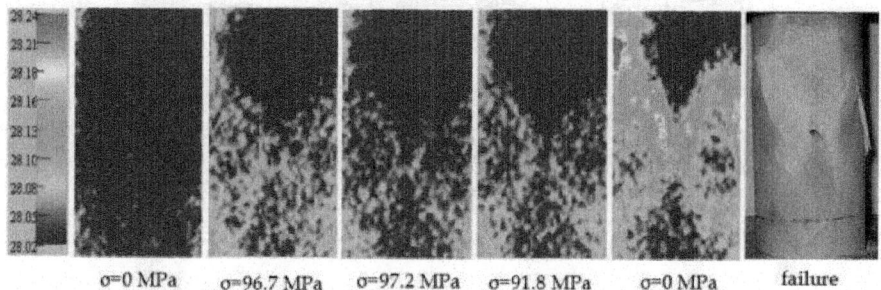

σ=0 MPa   σ=96.7 MPa   σ=97.2 MPa   σ=91.8 MPa   σ=0 MPa   failure

Figure 2. The IRR image positive anomaly of "X"-shaped shearing fracturing of an uni-axially loaded marble sample.

The "//"-shaped positive IRR abnormal strips generated with loading along the "//"-shaped shearing zone at the upper part of sample before peak stress, and got distinguished after peak stress, as in Fig 3. The evolution of the positive IRR image anomaly had also reflected the fracturing being not symmetrical in that the upper part of the IRR anomaly strip was clear with higher temperature, while the lower part was fuzzy excepting for the final fracturing near the bottom of the sample. Besides, there was strong IRR anomaly spot at the fracturing center for the intensive accumulation of mechanical energy and for the intensive generation of frictional thermal at the local central place.

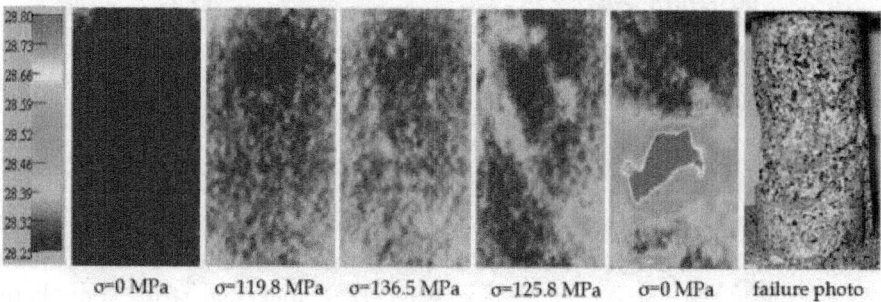

σ=0 MPa   σ=119.8 MPa   σ=136.5 MPa   σ=125.8 MPa   σ=0 MPa   failure photo

Figure 3. The IRR image positive anomaly of "//"-shaped shearing-fracturing of an uni-axially loaded granite sample.

The "|"-shaped negative IRR abnormal strip generated with loading along the tensile fracturing zone of a rock sample before the peak stress, and got distinguished gradually at the peak stress and after fracturing, as the approximately vertical dark strip in Figure 4. The same phenomenon for a sandstone sample with a calcite vein was also reported (Wu, et al., 2000).

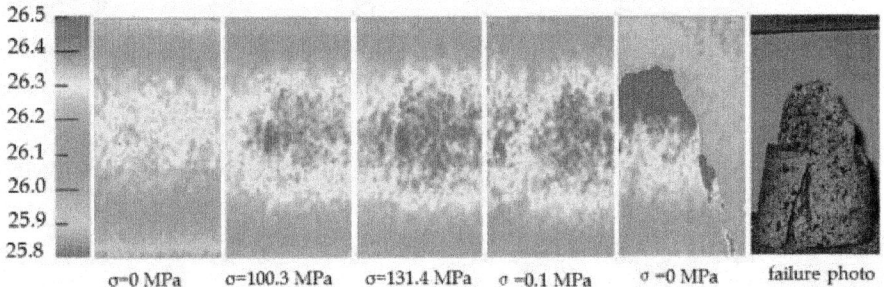

Figure 4. The IRR image negative anomaly of "|"-shaped tensile fracturing of an uni-axially loaded granite sample.

Uni-Axially Loaded Rock with a Central Hole

More than 10 samples with a central hole, modeling the structure stability of loaded rock tunnels, made from marble and granite were infrared imaging detected. The rock samples had two kinds of shapes respectively being cylinder with diameter and length, respectively, 50 and 100mm, and regular block with thickness, width and length, respectively, 70, 35 and 100 mm. It was discovered that there were distinguished positive IRR image anomalies before rock fracturing, and the place of anomaly were exactly the coming fracturing place. As in Figure 5, the positive IRR image anomalies had reflected the two kinds of fracturing, respectively being diagonal fracturing (sample 1~5) and fork fracturing (sample 6). The IRR anomalies, along the fracturing planes and shaped as spots or strips, generated not only on rock surface but also on the hole's surface (lateral sample 4 and 5). The temperature increment is 1~3°C and 4~8°C respectively for marble and granite samples.

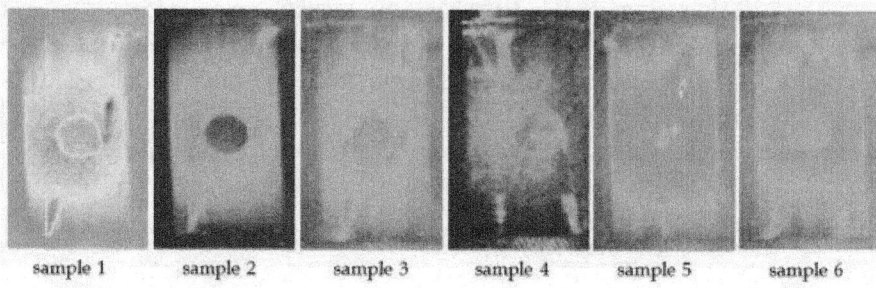

sample 1   sample 2   sample 3   sample 4   sample 5   sample 6

Figure 5. The IRR image positive anomalies of a group of uni-axially loaded marble samples with a central hole.

Compressively Sheared Rock

More than 20 samples, size $7 \times 7 \times 7 \times cm^3$, made from sandstone, marble, limestone, granite and gneiss, were compressively sheared and infrared imaging detected. Three pairs of steel platens with shearing angle being 45°, 60°and 70° respectively were applied. The loading rate was controlled as 2~5kN/s. It was discovered that the IRR temperature of rock surface changed with loading, and a strip-shaped positive IRR image anomaly generated along the central shearing plane before fracturing. With loading, the positive abnormal strip got more and more distinguished and migrated gradually from the upper end to the lower end of the sample, which foretold that the compressive-shearing fracturing was developing gradually from the upper end to the lower end of the sample along the central shear plane. Figure 6shows the typical IRR image series of a compressively sheared limestone sample.

As a special geological phenomenon occurring with the formation of great fault, penniform-shaped fractures are a group of secondary fractures produced with the formation of great primary fracture (Nicolas et al., 1977). It happened to occur in our experiments that there were penniform-shaped fractures produced with a primary fracture in the compressive loaded rock samples, as in Figure 7. The IRR positive anomaly strips generated aside the primary IRR strip, passing through the central shearing plane, had reflected the penniform-shaped fracturing events.

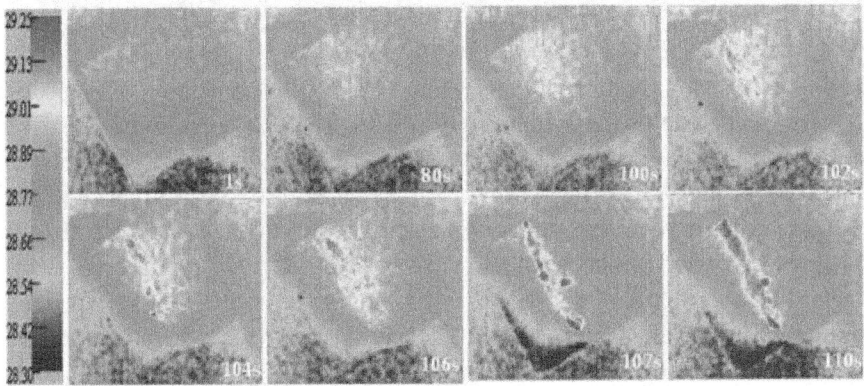

Figure 6. The IRR image positive anomaly of the fracturing of a compressively sheared limestone sample (time in second).

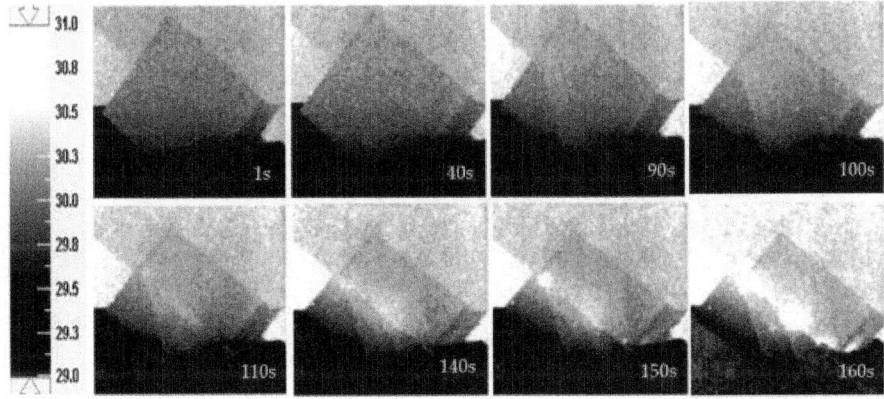

Figure 7. The IRR image anomaly of the penniform-shaped fracturing of a compressively sheared marble sample.

Bi-Sheared Frictional Sliding Rock Blocks

Ten groups of rock samples made from gabbro, granodiorite, limestone and marble were IRR detected in the process of bi-sheared frictional sliding or viscosity sliding. Each group, as in Figure 8 and 9, was comprised of three jointed rock blocks whose size respectively be $50 \times 50 \times 100mm^3$, $50 \times 70 \times 150mm^3$ and $50 \times 50 \times 100mm^3$ from left to right, and its friction area was constant, $50 \times 100mm^2$. Four contact conditions, symmetrical (yes for rock property and for its smooth friction surface, as in Figure 8), uncertain symmetrical (yes for rock property but not for its coarse friction surface, as in Figure 9), unstable asymmetrical (yes for rock property but not for its staged friction surface) and stable asymmetrical (not for rock property but yes for its smooth friction surface), were designed and tested respectively (Wu et al., 2004b).

It is revealed that the evolution of rock surface IRR temperature field is not only correlated with rock stress, but also correlated with the features of friction surface and rock properties at both sides. General law lies in that the IRR at the place of stress concentration and strong friction zone is stronger than that at the place of stress relaxation and weak friction zone. In condition of friction surface be symmetrical, the IRR image is double butterfly-wings shaped, as in Figure 8. However, in condition of friction surface be uncertain symmetrical, unstable asymmetrical or stable asymmetrical, the temporal-spatial evolution of IRR anomaly is uncertain or unstable, as in Figure 9. The positive IRR anomaly spots, foretelling the evolution of stress, energy and viscosity-sliding process, may be beads-shaped, needle-shaped, suspended needle-shaped, strip-shaped, single butterfly-wings shaped or its evolution in order (Wu et al., 2004b).

Figure 8. The IRR image positive anomaly of the stick-slipping of symmetrical rock samples (time in second).

Figure 9. The IRR image positive anomaly of the stick-slipping of asymmetrical rock samples (time in second).

Bi-Axially Loaded Rock

Infrared imaging detection on the rupturing of en echelon and collinearly disjointed jointed faults were done in the process of bi-axial loading. It was revealed that the IRR from loaded rock surface is correlated with loading stress, which could be divided into five stages as loading beginning, linear elastic, stress locking, stress unlocking and fracturing(Wu et al., 2004a). During the stress-unlocking stage, positive IRR anomaly strip generated at the disjointed zone, as in Figure 10 and 11. The positive IRR anomaly strip around the disjointed zone has general evolution features as: firstly, the strip gets enhancing; then, gets weakening (or 'silence'); and finally, gets enhancing again. The re-enhancing of IRR anomaly strip after the weakening stage is a meaningful precursor foretelling the place of primary fracturing of faults or the coming epicenter of an EQ.

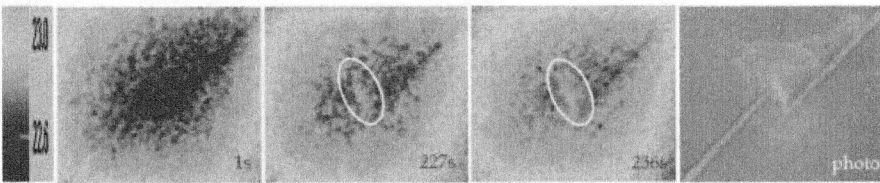

Figure 10. The local IRR positive anomaly foretell the fracturing of en echelon disjointed faults (marble, time in second).

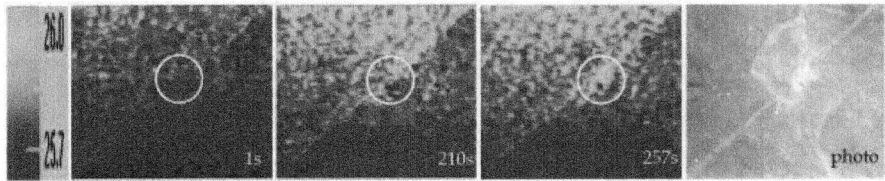

Figure 11. The local IRR positive anomaly foretell the fracturing of collinearly disjointed faults (marble, time in second).

Rock Fracturing Precursor: IRR Temperature Anomaly

Quantitative Index: AIRT

The evolution of surface IRR from loaded rock is the comprehensive effect of rock thermo-elastic acting, pore gas desorbing & escaping, fractures producing & extending, rock frictionating, heat transferring and environment radiation. Being the integral reflection of surface IRR energy, the average IRR temperature (AIRT) is selected as a quantitative index to study the evolution of IRR from loaded rock and to seek for rock fracturing precursors (Wu et al., 2006b). The infrared imager detects and records the thermal images of loaded rock surface. The thermogram is comprised of a matrix of color pixels which representing the IRR brightness temperature of each pixel of the rock surface. For example, the imaging matrix of TVS-8100MKII infrared imager is 160×120. The IRR temperature of each pixel tends to fluctuate with time due to the instability of the detector unit and the influence of environmental radiation, and the IRR temperature of each pixel will not always be the same for the local difference of rock stress and rock strain. The maximum, minimum and average value of loaded rock surface IRR temperature, respectively being IRRT$_{max}$, IRRT$_{min}$ and IRRT$_{ave}$, could be quantitatively obtained from thermogram. The analysis revealed that IRRT$_{max}$ and IRRT$_{min}$ will not change obviously except that IRRT$_{max}$ might rise suddenly just before rock fracturing, while IRRT$_{ave}$ is to change stably with loading, as in Figure 12. The physical interpretation lies in that the surface IRRT$_{ave}$ is a general reflection of the energy balance inside the loaded rock.

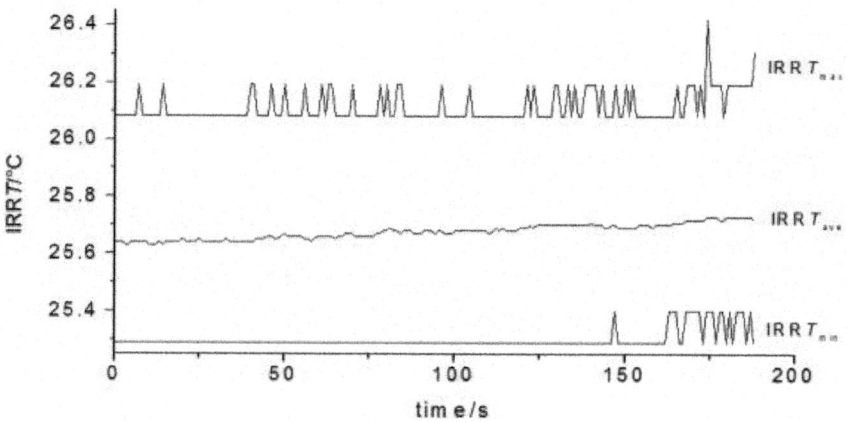

Figure 12. The evolution of three indexes of IRR brightness temperature of loaded rock surface.

Hence, the IRRT$_{ave}$ of rock surface, denominated as AIRT, is selected as a quantitative index to study the precursors of rock fracturing and geo-hazards. The procedures for AIRT-based precursor analysis includes: 1) to define a unified boundary of the analyzed region (resampling region) for all thermogrames; 2) to resample the IRR temperature value from the data file of each thermograme in time order; 3) to calculate the AIRT of the resampling region of each thermograme; 4) to draw the AIRT–time curve of the rock sample; 5) to analyze the evaluation features of the AIRT–time curve and to identify the messages as a precursor of rock fracturing and hazard; 6) to compare with the qualitative image anomaly so as to analyze and to confirm the AIRT abnormal precursor.

Influence Factors of AIRT Curves

Loading stages and rock deformation

The stress-strain curve is a basic method for describing rock deformation and for interpreting rock mechanical behaviors (Hudson and Harrison, 1997). Generally, the deformation process of loaded rock is divided into four stages respectively being stage-I of defects compaction, stage-II of linear elastic deformation, stage-III of plastic deformation and stage-IV of fracturing failure, as in Figure 13. The four characteristic points, E, Y, P and F are called as elastic-starting point, yield-starting point, peak-stress point and failure-impending point respectively.

Stage-I: the downward-concave curve section tells that there are some defects such as pores, fissures and joints inside the rock body, and that the defects is under compaction, which cause the stress to rise slowly. The more the defects, the severe the curve downward concave.

Stage-II: the curve section linearly developed tells that the compaction of defects has finished and the rock is undergoing elastic deformation. The higher the angle of the section line inclined, the stronger the rock.

Stage-III: the upward-concave curve section tells that there are new fractures developing inside the rock. The plastic deformation starts, and the new generated fractures together with the initial defects are possible to cause friction between its two side-faces.

Stage-IV: the curve section turning to drop tells that the fractures are getting wider, longer and to connect with each other. The rock is losing its strength and stability, and the final fracturing failure or rock hazard is impending.

For the difference between rock compositions, the details of stress-strain curve of different rock will be different. As to brittle rock, its stage-II is close to point P and its stage-IV will be cliff-shaped. Usually, most of the crust rocks are brittle. Five kinds of typical crust rock, granodiorite, gabbro, gneiss, limestone and marble had been tested in our experiments. The typical load-displacement curves of the tested rock samples are shown in Figure 14. It tells that all the tested rocks are brittle.

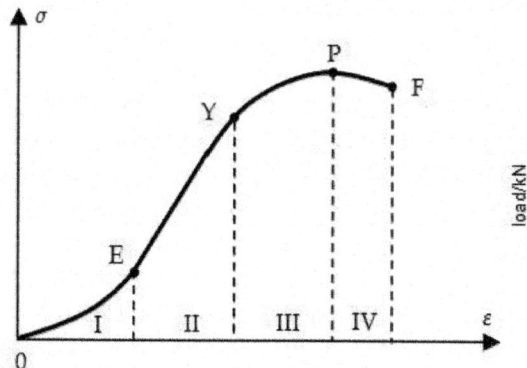

Figure 13. Typical AIRT curve of uni-axially loaded rock.

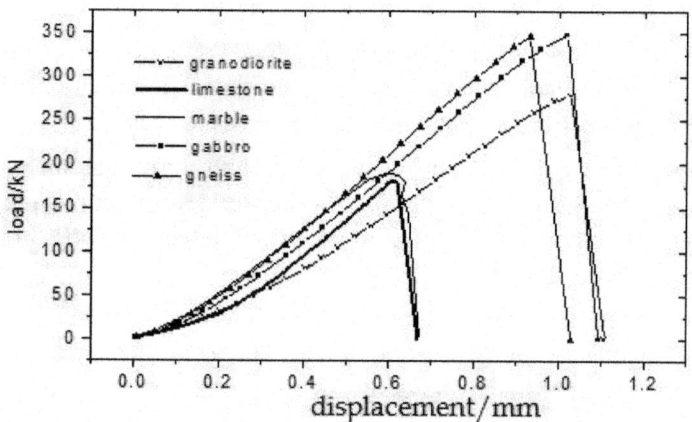

Figure 14. The load-displacement curves of five tested rocks.

The influence of loading condition

The experiments discovered that the evolutions of AIRT have different laws in different loading condition, such as uni-axial loading (constant displacement controlled), compressively-sheared loading (approximately constant load controlled) and bi-axial loading (constant displacement controlled).

Uni-axial loading

As shown in the left-hand side of Figure 15, the surface facing to the infrared imager is to be detected, and a rectangle region close to the boundary of the rock sample is defined for data resampling and analyzing (Liu et al., 2002; Wu et al., 2002). Multiple experiments revealed that there was slight variation of AIRT at different deformation stage although the AIRT linearly increased with load and deformation. At stage-I, the AIRT will rise slowly or drop a little; at stage-II, the AIRT will rise stably; at stage-III, the AIRT will rise quickly than that in stage-I and stage-II. The right-hand side of Figure 15shows the comparison of the evolution of AIRT and the load with rock deformation, which is rock displacement in generally, of a marble sample.

Compressive-shear loading

In condition of compressively shear, the rock sample will always get fracturing along the shearing plane, which locates near to the central plane of the loaded sample. To minimize data resampling work and to focus on the key region, a narrow rectangle along the shearing plane is defined as the resampling and analyzing region (Wu et al., 2004c). Multiple experiments revealed that the temporal evolution of the AIRT is different with the shearing angle. Three shearing angles (γ) being 45°, 60° and 70°respectively, are applied. As the shearing angle changes from 45° to 70°, the temporal evolution changes from monotonic rise, to drop-to-rise and to monotonic drop in order, as in Figure 16.

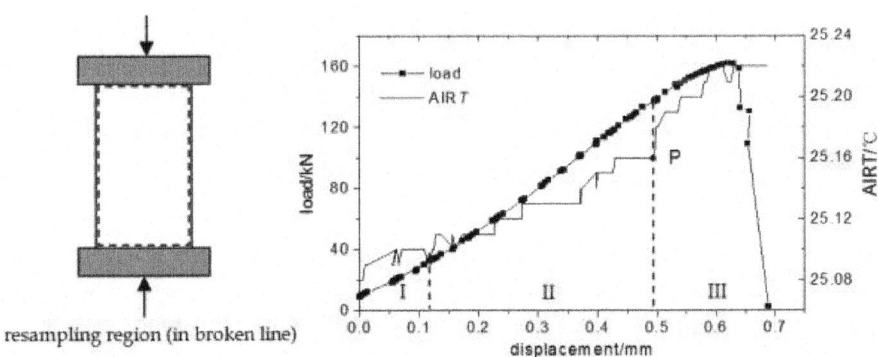

Figure 15. The typical AIRT curve of uni-axially loaded rock sample (marble).

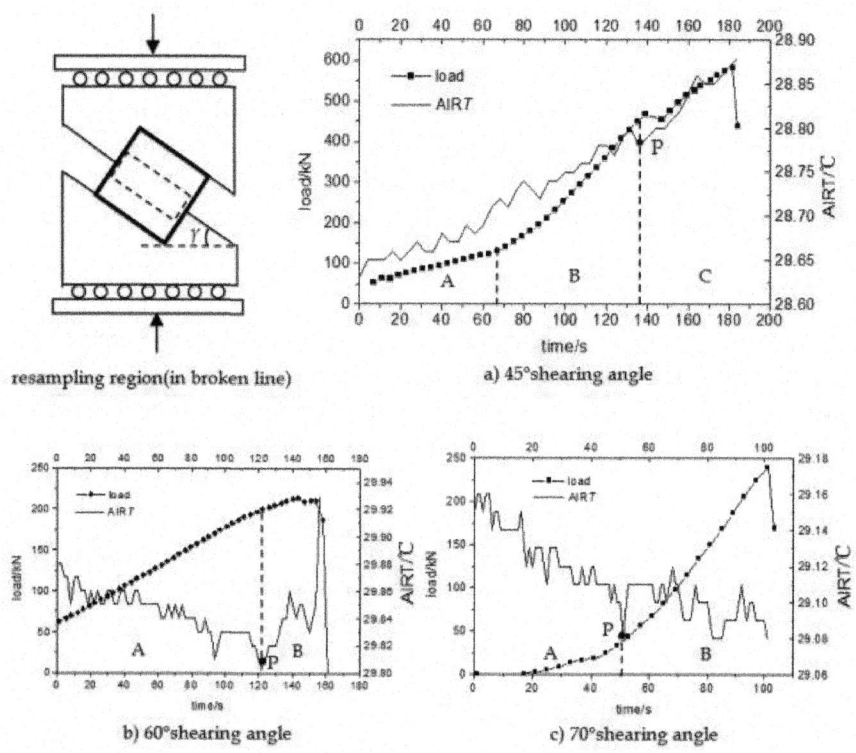

resampling region(in broken line)

a) 45°shearing angle

b) 60°shearing angle

c) 70°shearing angle

Figure 16. The typical AIRT curve of compressively-sheared rock samples at different shearing angles.

The mechanism lays in that the ratio of compressive-stress to shear-stress along the shearing plane decrease with the rise of the shearing angle. The smaller is the shearing angle, the higher is the compression-shear ratio. In condition of 45°, the load-time curve developed in three stages, as stage A, B and C in Figure 16a, with load increasing from slow to rapid, and to slow again. In condition of 60°the load-time curve developed in two stages, as stage A and B in Figure 16b, with load speed changing from approximate constant to be decrease slightly. In condition of 70°, the load-time curve developed in two stages, as stage A and B in Figure 16c, with loading speed changing from slow to rapid.

The compressive action on loaded rock is to cause surface IRR temperature rise, while the tensile action on loaded rock is to cause surface IRR temperature drop. Actually, both compressive action and tensile action are to occur along the compressively sheared plane, and the detected surface IRR is the comprehensive effect of the two actions. It was reached that (Wu et al., 2004c): 1) in condition of shearing angle being 45°, the surface AIRT will rise monotonically with loading in that the temperature increment from compressive action and friction is stronger than the temperature decrement from tensile action in the whole loading process; 2) in condition of shearing angle being 60°, the surface AIRT will drop monotonically with loading in that the temperature increment from compressive action and friction is weaker than the temperature decrement from tensile action before stage-III (point Y in Figure 13, and point P in Figure 16b); with the friction effect getting strong in stage-III, the surface AIRT will get to rise in that the temperature from compression and friction get stronger than the temperature decrement from tensile action; 3) in condition of 70°, the surface AIRT will drop monotonically with loading in that the temperature increment from both compression and friction are weaker than the temperature decrement from tensile action.

Biaxial loading

By using of bi-axial loading system and infrared imaging system, the IRR features of two kinds of disjointed jointed faults, respectively be collinearly and non-collinearly disjointed faults, were experimentally studied (Wu et al., 2004a). Since all the faults got fractured finally at the disjointed zone, a circle covering the disjointed region is defined as the resampling region, as in the left-hand side of Figure 17. It could be known from the right-hand side of Figure 17 that the IRR from loaded samples is related with load stress, and the evolution stage could be classified into five stages (I~V) relating with initial compacting, elastic deforming, stress blocking, stress deblocking and rock fracturing respectively. From stage-II to stage-IV, the evolution of AIRT has the features of rising to dropping, and to rising again.

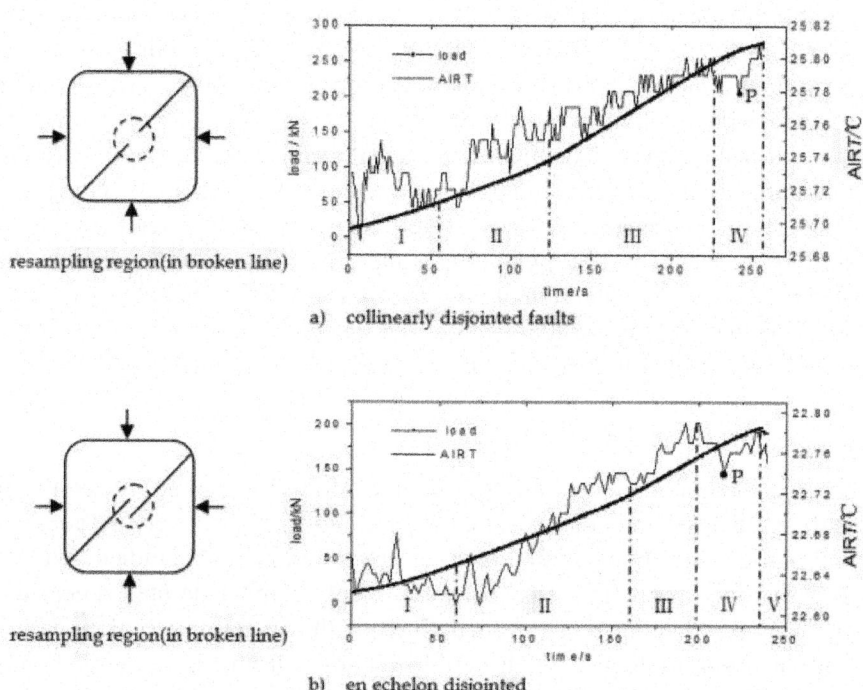

Figure 17. The typical AIRT evolution of two kinds of disjointed faults bi-axially loaded.

The influence of rock characteristics

It was discovered in our experiments that for the most of rock samples in condition of uni-axial loading, its AIRT approximately rose with loading. But there were a few abnormal samples made from limestone had shown AIRT features of dropping with loading, as in Figure 18. The cause lies in that limestone has much more pores than the other rocks. Usually, there are many gases, such as CH_4, CO_2, CO and O_2etc., enclosed inside the pores of rock body (Wang, 2003). With the decrement of pore volume due to the loading compaction and with the increment of fractures produced inside the limestone sample, the pore gases will get escaping. The escaping behavior of pore gas needs to absorb thermal energy from the rock sample. If the heat from compression and friction is lower than that absorbed by pore gases, the surface AIRT is to drop with loading. If look carefully at the load-displacement curves in Figure 14, it could be founded that the curve of limestone concaved downward the most at the compaction stage as compared to that of the other four kinds of rock, which means that there are more pores inside limestone than the others.

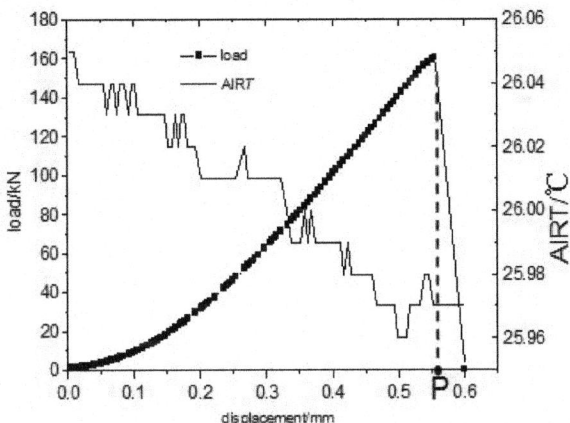

Figure 18. AIRT of uni-axially loaded limestone sample.

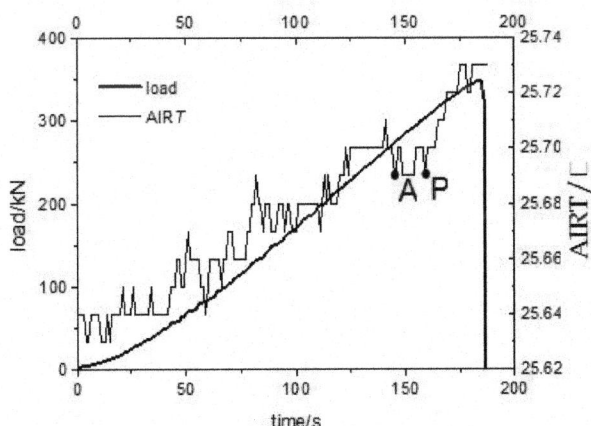

Figure 19. AIRT short-dropping precursor for uni-axially loaded gabbro sample.

The Classification of Precursors

Analysis to the evolution of AIRT curves discovered that a large amount of rock samples in condition of uni-axial loading, compressively-sheared loading and bi-axial loading had presented obvious precursors for rock fracturing. As referring to the general process of AIRT evolution, the AIRT anomaly precursors for rock fracturing and hazard could be classified as short-dropping, rapid-rising and dropping-to-rising respectively.

Short-dropping precursor

The AIRT curve rises with loading but has a short dropping at loading stage-IV; later, the AIRT curve will rise again. The bi-axially load on collinearly and non-collinearly disjointed faults had shown short-dropping precursors as in Figure 17, and the point P was suggested to be the precursor point of rock fracturing and rock hazard. Figure 19 shows another typical case of gabbro sample uni-axially loaded. Here, point A is the turning point of AIRT from rising to short dropping, and point P is another turning point from short dropping to rising again, which is suggested to be the precursor point of rock fracturing and rock hazard.

Rapid-rising precursor

The AIRT curve rises slowly with loading but turns to rise rapidly before rock fracturing, and the turning point is exactly the precursor point. Figure 15 and Figure 16a have this kind of precursor. Figure 20 shows another typical case of marble sample uni-axially loaded. Here, point P is the turning point of AIRT from rising slowly to rising fast, which is suggested to be the precursor point of rock fracturing and rock hazard.

Dropping-to-rising precursor

The AIRT curve drops slowly with loading but turns to rise just before rock fracturing, and the turning point is exactly the precursor point. Figure 16b has this kind of precursor. Figure 21 shows another typical case of marble sample uni-axially loaded. Here, point P is the turning point of AIRT from dropping slowly to rising fast, which is suggested to be the precursor point of rock fracturing and rock hazard.

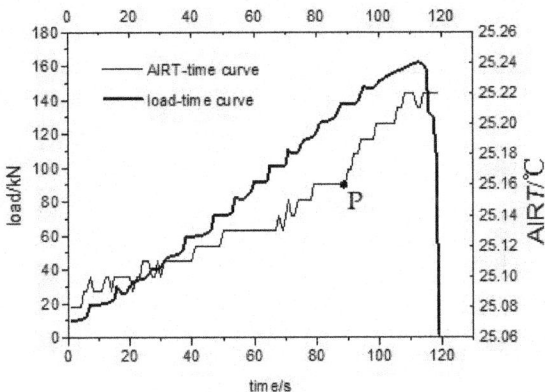

Figure 20. AIRT fast rising precursor for uni-axially loaded marble sample.

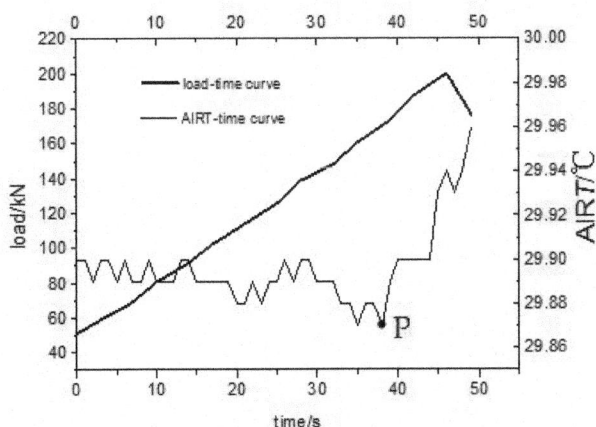

Figure 21. AIRT dropping-to-rising precursor for compressively sheared loaded marble sample.

The Temporal Features of Precursors

The occurrence moments of the precursors of AIRT of totally 52 tested rock samples are listed in Table 1(Wu et al, 2006b). Although the loading conditions and the rock samples are different, the precursor occurrence moment, were very similar as $0.77{\sim}0.94\sigma\,c\,(\sigma_p)$. Here, σ_c is the uni-axial compressive strength, and σ_p is the peak stress. The precursor occurrence moment of uni-axially loaded or compressive sheared rock sample is $0.79\sigma_c$ and $0.82\sigma_c$ respectively. It is worthy to mention that the precursor occurrence moment of bi-axially loaded collinearly disjointed faults and echelon faults are much different. That for collinearly disjointed faults was close to the peak stress, $0.87\sigma_p$, while that for echelon faults was far away from peak stress, $0.77\sigma_p$. It provides an important evidence for the complexity of study on tectonic EQ prediction on shock time, based on satellite infrared remote sensing and referring to the seismogenic mechanism.

Table 1. Statistic for precursor occurrence in different loading conditions.

Loading condition		Sum of tested samples, St	Sum of samples with precursors, Sp	The ratio: (Sp/St)×100%		Average of precursor occurrence moment
uni-axial loading		22	9	41%		$0.79\,\sigma_c$
compressively-sheared loading	70°	7	1	14%	58%	$0.94\sigma_p$
	60°	8	6	74%		$0.82\sigma_p$
	45°	11	8	73%		$0.77\sigma_p$
Bi-axial loading for collinearly disjointed faults		2	2	100%		$0.87\,\sigma_p$
Bi-axial loading for en echelon faults		2	2	100%		$0.77\,\sigma_p$

Large IRR at Fracturing Centre

As the recording rate of the infrared imager applied was 60f/s, the transient IRR temperature at the fracturing center could be snapped. In condition of uni-axial loading, the fracturing center of a brittle rock is usually at the "X"-shaped fracturing center. It is discovered that the transient IRR temperature at the fracturing center is much higher than that on rock surface, as in Figure 22, and it is positively related with rock strength and rock deformation. For some compressively sheared hard rock samples made from gabbro and gneiss, the transient IRR temperature at the fracturing center is higher than 155°C, which is the upper limitation of the 2nd temperature range (72~155°C) of the imager applied.

In condition of high angle, 60°and 70°, compressively sheared loading, the fracturing center is at the center of the fractured shearing zone. Since the ruptured upper block of rock sample was pushed apart from the steel platen immediately after the abrupt rupturing, usually be 1~2 s after the rupturing, the inside shearing zone got exposed to the imager immediately and the transient IRR temperature filed was snapped. It was discovered that the IRR temperature on the inside shearing zone is not only much higher than that of outside rock surface, but also inhomogeneous distributed, neither even nor centripetal, as in Figure 23. It means that much more mechanical energy had been converted into frictional thermal and IRR energy due to the intensive energy accumulation, the sufficient local deformation and the abrupt frictional sliding at the center of the shearing zone. In other words, the large IRR temperature at the inside shearing center had reflected the comprehensive effect of local concentrated energy conversion and frictional thermal.

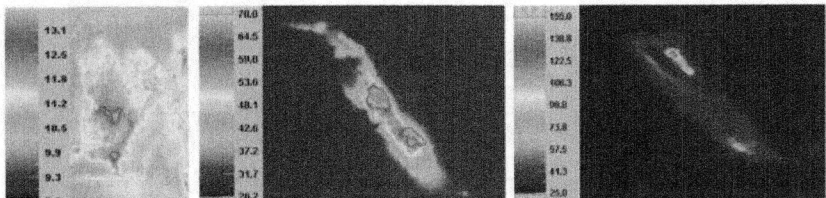

a) uni-axially loaded granite b) compressively sheared gabbro-1 c) compressively sheared gabbro-2

Figure 22. The transient IRR thermogram of fracturing rock samples.

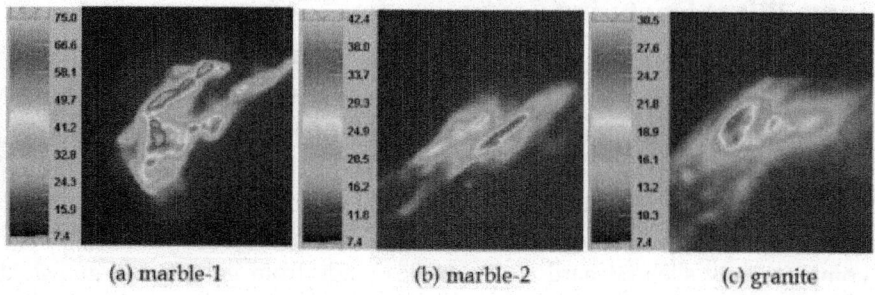

| (a) marble-1 | (b) marble-2 | (c) granite |

Figure 23. The inside IRR isothermal filed on the fracturing zone of compressively sheared rock samples.

Hence, it could be deduced that in condition of great tectonic stress, large deformation and/or abrupt frictional sliding, large temperature as high as hundreds or thousands of degree Celsius is possible inside crust rocks. The high temperature could cause partial melt of crust rock, which provides a scientific explanation for the existence of pseudotachylyte in some lager faults (Nicolas et al., 1977, Sibson et al, 1980) and for the failure-generated EQ lights (Martelli et al., 1989). Besides, we can deduce that the continuous shearing deformation or the abrupt fracturing of highly loaded rock/coal body in a coal mine is possible to cause local sheared heating of temperature hundreds of degree Celsius, which might be a potential ignition of local methane (the minimum ignition temperature is 595°C).

REMOTE SENSING ROCK MECHANICS MODEL (RSRM-MODEL)

Thermo-Mechanical Coupling Effect

Thermo-Mechanical Coupling in a Loaded Solid

The heat production inside a loaded solid is called as thermo-mechanical coupling effect. According to the material features and the different deformation stages of a loaded solid, the thermo-mechanical coupling is classified as thermo-elastic, thermo-plastic and thermo-viscous respectively for elastic deformation, plastic deformation and viscous deformation. Generally the rock is a hard brittle solid, its plastic and viscous deformation could be ignored, and the thermo-elastic effect and the frictional thermal are the two chief mechanisms of surface IRR from loaded rock. Kelvin coined the thermo-elastic theory in 1853 that the changed physical temperature of a loaded component is correlated to its changed stress as follows:

$$\Delta T / T = -K_0 \Delta \sigma \tag{1}$$

Here: T is the absolute temperature of a loaded component (K); ΔT is the changed temperature (K); K_0 is the thermo-elastic factor (MPa^{-1}); and Δ_σ is the changed sum of three principal stresses ($\sum \sigma_i, i = 1,2,3$, MPa).

As for an isotropic linear elastic solid loaded bi-axially with a free surface, the surface physical temperature variation is tightly correlated with the sum of two principal stresses ($\sum \sigma_i, i = 1,2$):

$$\Delta T = -\alpha / \rho C_p \cdot [T \cdot \Delta(\sigma_1 + \sigma_2)] \tag{2}$$

Here: T is the surface absolute temperature of a loaded solid (K); ΔT is the changed temperature (K); α is the factor of linear expansion (K^{-1}); ρ is the solid density (Kg m^{-3}); Cp is thermal capacity of solid at normal atmosphere (J Kg^{-1} K^{-1}); σ_1 and σ_2 are the two principal stresses (MPa). The thermo-elastic factor K is defined as $K=-\alpha/\rho C_p$.

For the mechanism of stress measurement with TSA and SPATE, the relationship between the stress increment and the IRR signal based on equation (2) is as follows (Mounatin and Webber, 1978):

$$\Delta(\sigma_1 + \sigma_2) = A_{th} \cdot \Delta S$$
$$\Delta S = \Delta(\sigma_1 + \sigma_2) \cdot A_{th}^{-1} \cdot \tag{3}$$

Here: A_{th} is a comprehensive factor called as corrective factor, which is a function of solid surface emissivity, solid surface physical temperature, solid thermo-elastic factor and three parameters related to the IRR detector, unit in MPa U^{-1}. ΔS is the increment of thermo-elastic voltage signal detected (U).

Changed IRR Temperature of Loaded Rock

If the slight change of rock surface emissivity, rock thermo-elastic factor and the physical parameters of IRR detector during rock loading could be ignored, and if the changed rock surface physical temperature cannot be ignored due to the thermal exchange and the frictional thermal, the relationship between A_{th} and changed rock surface physical temperature could be expressed as $A_{th}=\beta \cdot T^{-1}$. The detected IRR signal S is a direct representation of surface IRR temperature, i.e., $\Delta IRRT=\gamma \cdot \Delta S$. Hence, the following equation for changed IRR temperature could be deduced:

$$\Delta IRRT = \gamma \cdot \beta^{-1} \cdot T \cdot \Delta(\sigma_1 + \sigma_2) \tag{4}$$

Here: $\Delta IRRT$ is the changed IRR temperature (K); β is a constant correction factor related to rock surface's emisivity, rock thermo-elastic factor and three parameters of IRR detector, in unit MPa K U^{-1}; γ is a transfer factor between detected voltage signal and IRR temperature (K U^{-1}).

It means in equation (4) that the changed IRR temperature of rock surface is a direct reflection of the changed sum of the two principal stresses. If no frictional thermal produced and the thermal exchange is stable, the IRR temperature of rock surface is to rise with loading, and the spatial-temporal evolution of surface IRR image will be stable. If there is no frictional thermal produced but the thermal exchange is unstable, the surface IRR temperature will be unstable, and the spatial-temporal evolution of surface IRR image will also be unstable. If there is frictional thermal produced inside and conducted to rock surface, both the thermal exchange and the surface IRR temperature of rock surface will be unstable, and the spatial-temporal evolution of surface IRR image will get complicated, as in Figures 2~5, 10, 11.

Especially, in condition of compressive shearing and fictional sliding, there is a large amount of frictional thermal produced in the friction zone, which is to cause the rise of physical temperature in friction zone. The rock surface temperature will rise if the thermal conduction from the friction zone can reach to rock surface. Hence, the IRR image anomaly will be distinguished and sometimes be large as a combined effect of rock stress and frictional thermal, as in Figures 6~9.

Remote Sensing Rock Mechanics Model

Energy-related IRR from Loaded Rock

As a relatively independent closed system comprised of loader head, rock sample and environmental air, as in Figure 24, the rock deformation, rock fracturing and rock hazard are all of a complex process of energy input and consumption. If the possible chemical reactions inside a loaded rock can be ignored, the inputted energy of a loaded rock will include the mechanical work from loader and the heat input through positive thermal exchange from loader head and environmental air. The energy consumption by the loaded rock is much more complex including the energy accumulation in rock and the energy dissipation from rock.

The energy accumulation in a loaded rock includes the positive elastic-plastic deformation energy of rock (the positive change of oscillation and rotation energy of mineral molecules), the surface energy of new produced fractures or fissures, and the friction actions between mineral molecules, grains, joints, fissures and fractures inside the rock as well as thus produced frictional thermal. The energy dissipation from loaded rock includes the negative thermal exchange with the loader head and/or environmental air (i.e., heat output), the kinetic energy of departed fragments of fractured rock, the light radiation, acoustic emission, radio frequency emission and IR & microwave radiation.

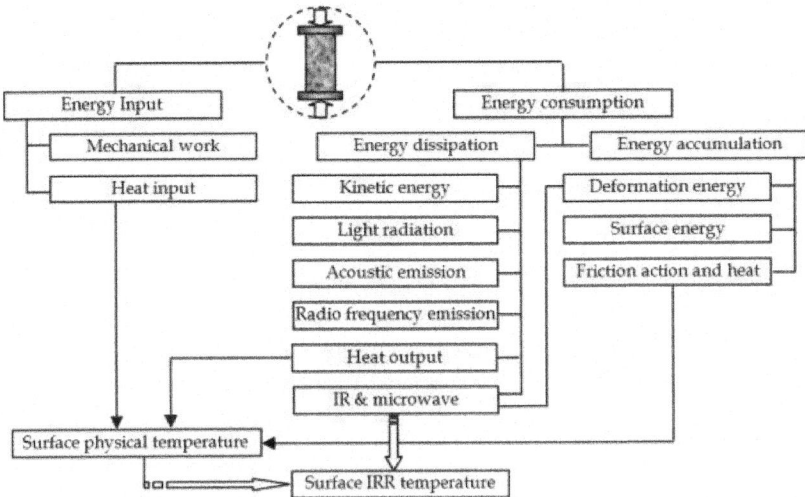

Figure 24. The IRR mechanism related to the energy accumulation and consumption of a loaded rock.

The thermal exchange and the friction action are to change the heat state of a loaded rock, and the rock surface physical temperature is a direct index reflecting the heat state of the loaded rock. Stephen-Boltzmann law states that the IRR strength (radiation flux density) of any material, at temperature above absolute zero degree, is biquadratic to its surface physical temperature. Crystal Physics states that the energy jump of molecules oscillation and/or rotation due to the change of molecules distance, resulting from deformation, is an important mechanism of electromagnetic radiation. Hence, rock surface IRR is a comprehensive effect of rock deformation and rock surface thermal state. Rock surface IRR temperature could be a detective index reflecting rock surface physical temperature and rock surface deformation field, which implicating the complex physical-mechanical process inside the loaded rock.

In spite of thermal exchange and plastic deformation, the thermo-elastic effect and the frictional thermal are two of the main mechanics of changed IRR from loaded brittle rock. In the stage of elastic deformation, the thermo-elastic effect is the main cause; while in the stage of plastic deformation or fracturing, the friction-thermal effect plays a great role. At the moment of rock fracturing or hazard, the fraction-heat effect gets more distinguished. The frictional thermal effect depends on two factors being frictional force (decided by normal stress and frictional coefficient) and frictional speed respectively. The larger the frictional force and the quicker the frictional speed, the more the frictional heat.

RSRM-Model Based on Independent System

The rock sample, load header and environment air could be taken as a closed independent system, as in Figure 25, and the energy of the loaded rock sample is in a balance state as follows:

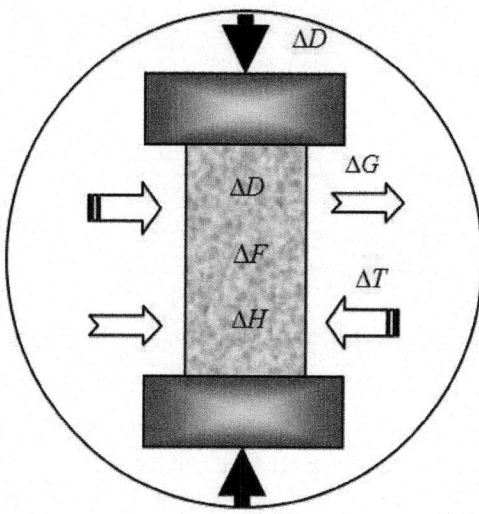

Figure 25. The energy balance of a loaded rock sample in a closed independent system.

$$\Delta M + \Delta T = \Delta D + \Delta F + \Delta H + \Delta G \qquad (5)$$

Here:

ΔM The inputted mechanical energy from loader, J; be positive;

ΔT The inputted thermal energy from loader and environment air, J; be positive;

ΔD The produced deformation energy of rock sample in elastic and plastic state, J; be positive;

ΔF The sum of consumed rock fracturing energy and formed fracture surface energy, J; be positive;

ΔH The heat energy increment of rock decided by its physical temperature, J; be positive if temperature rise or be negative if temperature drop;

ΔG The energy consumed by the desorbing and escaping of pore gas in rock samples, J; be positive.

From equation (5) we have:

$$\Delta H = (\Delta M + \Delta T) - (\Delta D + \Delta F + \Delta G) \tag{6}$$

The change of heat energy will result in the change of physical temperature, surface radiation energy and IRR temperature of loaded rock samples. Referring to Stephen-Boltzmann law, the AIRT is a direct index of rock radiation energy, and the change of AIRT (ΔAIRT) must have certain a relationship with the physical temperature (T) of loaded rock sample as:

$$E_{IR} = f(\text{AIRT}) = \varepsilon\sigma T^4 \tag{7}$$

Here:

E_{IR} The radiation energy of a loaded rock, J;

AIRT   The surface average IRR temperature of a loaded rock, K;

E   The radiation factor of rock sample, $0 < \varepsilon < 1$;

Σ The constant of Stephen-Boltzmann, $\sigma = 5.6679 \times 10^{-8}$, $J \times m^{-2} \times K^{-4}$;

Thermo-elastic effect is the basic mechanism of changed IRR from a loaded rock sample. Moreover, the desorbing and the escaping of pore gas, the expanding of initial fissures or joints, the friction between fissures, fractures, joints and grains, the thermal transfer between the rock, loader header and environment air, and the radiation from the environment all are to have thermal effect on loaded rock samples. The thermal state is comprehensively affected by the six factors which are rock stress, pore gas desorbing & escaping, rock fracturing, heat transferring, rock frictionating and environment radiation respectively. The equation could be expressed as follows:

$$\Delta AIRT = f[\Delta(\sigma_1 + \sigma_2), \Delta G, \Delta F_1, \Delta H, \Delta F_2, \Delta E]$$
$$= f_1(t) + f_2(t) + f_3(t) + f_4(t) + f_5(t) + f_6(t) \tag{8}$$

Here:

ΔAIRT   The detected change of AIRT, K; be positive if rise, be negative if drop;

$f_1(t)$ IRR temperature change due to thermo-elastic effect ($\Delta(\sigma_1+\sigma_2)$, K; be positive or negative;

$f_2(t)$ IRR temperature change due to pore gas adsorbing & escaping (ΔG), K; be negative;

$f_3(t)$ IRR temperature change due to the production of new fractures and the expansion of initial fissures, joints and new produced fractures (ΔF_1), K; be negative;

$f_4(t)$ IRR temperature change due to frictional thermal (ΔF_2), K; be positive;

$f_5(t)$ IRR temperature change due to heat transfer (ΔH), K; be positive or negative.

$f_6(t)$ IRR temperature change due to environment radiation (ΔE), K; be positive.

Thermo-elastic effect: $f_1(t)$

Referring to thermo-elastic theory and equation (4), $f_1(t)$ could be calculated as:

$$f_1(t) = \gamma \cdot \beta^{-1} \cdot T \cdot \Delta(\sigma_1 + \sigma_2)$$

(9)

In condition of uni-axially compressive loading, the load is to cause temperature rise in that σ_2 is constant zero and the positive σ_1 will linearly increase with loading. If $\Delta\sigma_1$ is positive, the $f_1(t)$ will be positive; If $\Delta\sigma_1$ is negative, the $f_1(t)$ will be negative. Hence, the $f_1(t)$ will be positive before the compressive stress peak, and will turn to be negative after the compressive stress peak.

In condition of uni-axially tensile loading, the load is to cause temperature drop in that σ_2 is constant zero and the negative σ_1 will linearly increase with loading. If $\Delta\sigma_1$ is positive, the $f_1(t)$ will be positive; If $\Delta\sigma_1$ is negative, the $f_1(t)$ will be negative. Hence, the $f_1(t)$ will be negative before the tensile stress peak, and will turn to be positive after the tensile stress peak.

In condition of compressively-sheared loading, the rock sample will be compressed by the normal component of the load. However, as to the central shearing plane, it will suffer not only compressive stress but also shearing stress. The σ_1 refers to the positive compressive stress normal to the shearing plane, while the σ_2 refers to the shearing stress which is actually negative tensile stress along the shearing plane. Hence, the ΔAIRT is decided by the sum of compressive stress and the tensile stress. If $\Delta(\sigma_1+\sigma_2) > 0$, the $f_1(t)$ will be positive; If $\Delta(\sigma_1+\sigma_2) < 0$, the $f_1(t)$ will be negative. As to Figure 1c,it's easy to know that $\Delta(\sigma_1+\sigma_2)$ will be positive if the shearing angle $\gamma 45^0$, $\Delta(\sigma_1+\sigma_2)$ will be zero if the shearing angle $\gamma=45^0$, $\Delta(\sigma_1+\sigma_2)$ will be negative if the shearing angle $\gamma 45^0$.

Pore gas desorbing & escaping effect: $f_2(t)$

Any rock has pores of different size more or less inside. Some rock, especially the sedimentary rock, may has certain gas, such as CH_4, CO_2, CO and O_2, enclosed in the pores or/and absorbed on the pore surface (Wang, 2003; Yang et al., 1999). Usually, most of the pores are enclosed and the gas molecules stay inside both in free gassy state and absorbed state. If the rock is loaded and suffers deformation, the volume of pores will decrease which results in the escape of gas from the pores. Once the load and deformation cause the enclosed pores getting fractured, the gassy molecules will escape firstly and the absorbed molecules will get desorbed to be gassy molecules and escape later. Both desorbing and escaping actions need to make use of heat energy from the rock, and thus will result in the AIRT drop of rock surface. Hence, the $f_2(t)$ is always negative, and the more the pores and gas enclosed, the more the negative effect of $f_2(t)$.

Fracture effect: $f_3(t)$

With loading and deforming, the rock is to get fractured. The new produced fractures together with the initial fissures and joints will extend both in width and length. The production of new fractures needs to consume energy, and the extension of fissures, joints and fractures also needs to consume energy. Hence, the $f_3(t)$ is always negative, and the more the fractures produced and fissures, joints and fractures extended, the more the negative effect of $f_3(t)$.

Frictional thermal effect: $f_4(t)$

With loading and deforming, the friction action is to occur between rock fissures, rock joints, rock grains, and new produced fractures. The friction action could be interpreted as: 1) at the beginning stage of loading, the friction may only be resulted from between rock fissures and between rock joints; 2) later, rock deformation increases with loading, new fractures are produced, and the friction between rock grains and between new produced fractures will join in; 3) finally, at the ending stage of loading, the rock deformation and fractures will be sufficiently developed, and the frictions between rock grains and between new produced fractures will be the chief contributors to the frictional thermal. In a word, the $f_4(t)$ is always positive no matter what is the principle friction factor. The more the friction, the more the positive effect of $f_4(t)$.

Heat transfer effect: $f_5(t)$

In the process of loading and inside the respectively independent loading system, the heat exchange is inevitable between the rock sample and the load header, the shearing platen or cushion-blocks, the surrounded atmosphere, etc. If the current temperature of rock sample is higher than the others, the heat of rock sample will be transferred out to whose temperature is lower. If the current temperature of rock sample is lower than the others, the heat of the others will be transferred into rock sample. Hence, the temperature of rock sample is a dynamic balance behavior between the heat transferred in and the heat transferred out. If the heat transferred in is more than that transferred out, the $f_5(t)$ will be positive; otherwise, it will be negative.

Environment-radiation effect: $f_6(t)$

The IRR detected by infrared imager includes not only the direct radiation from rock surface itself, but also the reflected radiation from environment. In laboratory, the chief environment radiations are the scattering sunshine, the moving human bodies and the illumination lamp. For the uncertain change of scattering sunshine, the movement of human bodies before the loaded rock sample, and the fluctuation of illumination light, the environment radiation effect on rock sample will be random. Hence, sometimes $f_6(t)$ may be positive, but sometimes $f_6(t)$ be negative. To eliminating the environment-radiation effect, the human bodies inside the laboratory were not permitted to move during testing process, the illumination lights were turned off, and the windows as well as its curtains were closed. Furthermore, some experiments were conducted in the evening so as to avoid the scattering sunshine completely.

Experiment Interpretation with RSRM-Model

Due to the comprehensive effects of $f_1(t) \sim f_6(t)$, the evolution of AIRT will be complex and will result in different possibility for abnormal AIRT precursors in different rock loading conditions, which including the loading scheme, rock type, and environment parameters. The following discussions are based on that $f_5(t)$ and $f_6(t)$ can be ignored.

Uni-axial Loading Experiments

For uni-axially loaded rock, $f_4(t)$ will take place only after that the rock has sufficiently deformed and the fractures have sufficiently developed.

At loading stage I and stage II, the rock surface AIRT are decided by $f_1(t)$ and $f_2(t)$. If the loaded rock is igneous rock or metamorphic rock, $f_2(t)$ is very rare since no gas absorbed in its pores usually, and the AIRT will rise with loading. If the loaded rock is sediment rock and with gas closed and absorbed in pores, $f_2(t)$ is inevitable, and the AIRT will rise if $f_1(t) > |f_2(t)|$, or be constant if $f_1(t)=|f_2(t)|$, or drop if $f_1(t) < |f_2(t)|$.

At loading stage III, fractures get sufficiently developed and pores get seriously damaged. The $f_3(t)$ begins to has more and more effect on evolution process of AIRT. AIRT will rise if $f_1(t) > |f_2(t)+f_3(t)|$, or be constant if $f_1(t)=|f_2(t)+f_3(t)|$, or drop if $f_1(t) < |f_2(t)+f_3(t)|$.

At loading stage IV, the friction action starts and $f_4(t)$ begins to have more and more effect on the evolution of AIRT. AIRT will rise if $[f_1(t)+f_4(t)] > |f_2(t)+f_3(t)|$, or be constant if $[f_1(t)+f_4(t)]=|f_2(t)+f_3(t)|$, or drop if $[f_1(t)+f_4(t)] < |f_2(t)+f_3(t)|$. Since $f_2(t)$ is very rare for igneous rock or metamorphic rock, the rise speed of AIRT of loaded igneous rock or metamorphic rock will get fast at stage IV, and the speed turning point is suggested to be the precursors point.

Compressive Shearing Experiments

For compressively sheared rock, not only $f_1(t)$ but also $f_4(t)$ is decided by shearing angle (γ). If $\gamma 45°$, $f_1(t)$ will be positive for $\Delta(\sigma_1+\sigma_2) > 0$, and the friction action will be much strong in that σ_1, which is normal to the friction plane, is large. If $\gamma=45°$, $f_1(t)$ will be zero since $\Delta(\sigma_1+\sigma_2)=0$. If $\gamma 45°$, $f_1(t)$ will be negative for $\Delta(\sigma_1+\sigma_2) < 0$, and the friction action will be much week since σ_1 is slight.

Biaxial Loading Experiments

For bi-axially loaded rock, $f_1(t)$ will always be positive. As to bi-axially loaded en echelon faults, collinearly and non-collinearly disjointed faults, $f_2(t),f_3(t)$ and $f_4(t)$ will occur simultaneously, and the evolution of AIRT will be fluctuated. If $[f_1(t)+f_4(t)] > |f_2(t)f_3(t)|$, AIRT will rise; if $[f_1(t)+f_4(t)] > |f_2(t)f_3(t)|$, AIRT will keep in the same level, and if $[f_1(t)+f_4(t)] < |f_2(t)f_3(t)|$, AIRTwill drop. Usually, the fact is that $[f_1(t)+f_4(t)] < |f_2(t)f_3(t)|$ at the fracturing stage, and there is a short drop of AIRT, which is called as 'silence' before EQ (Ohtake et al., 1981). However, $f_4(t)$ will be an important factor at the later stage of loading for the concentrated formation of fractures and friction in the disjointed zone, and the final state of AIRT will be rise for $[f_1(t)+f_4(t)] > |f_2(t)f_3(t)|$.

EARTHQUAKE THERMAL INFRARED ANOMALIES

The prediction of EQ is very difficult, but it's not impossible. A number of signs warning of EQs, such as foreshock activities, peculiar animal behaviour, increased low frequency EM-noise, concentrations of radon in water and air, ionosphere and magnetosphere perturbation, radio frequency emissions, terrestrial gas emanations, EQ clouds, and satellite TIR anomalies, have been proposed and reported during the past centuries. Satellite TIR anomaly was firstly reported in 1989, and had been repeatedly verified in the world during the past 20 years. It is becoming a prospecting space observation technology for seismic activity monitoring and for EQ predicating.

General Features of EQ TIR Anomaly

Gorny (1988) firstly reported that there were large area of TIR anomalies in METEO satellite remote sensing images, spatial resolution being 5 Km and wave length being 10.5~12.5μm, before many moderate-strong tectonic EQs in the mid-Asia and the east-Mediterranean region. Tronin (1996) analyzed 10000 about TIR images of channel AVHRR-2 of NOAA in ten years for the mid-Asia, and reached that there existed average anomaly, 1~5°C, before the EQs at this region, and that there was obvious statistical relations between the EQ and TIR anomaly. Qiang (1990), Cui (1999), Liu (1999),Xu (2000), Zhang(2002), Ouzounov(2004), Arun (2005), Liu (2007), Wu (2008) also reported that there occurred TIR anomalies in satellite images (NOAA, FY, MODIS) days before more than 100 EQs in Asia (China, India, Iran, Japan, Kamchatka, Pakistan, Turkey) and Europe (Italy, Greece, Spain). Analysis to all the reported cases, it was uncovered that satellite TIR anomaly before EQ has the following features generally (Wu et al., 2009):

- Temporal features: Satellite TIR anomaly usually appears 1~26 days before shock, and reaches to its peak 1~2 days before shock, and will disappeared soon after shock.
- Spatial features: The spatial distribution and geometrical shape of TIR anomaly is tightly related with tectonic structures such as plate borders and active faults. With the EQ impending, the TIR anomaly will move to or extend to gradually the coming epicentre along the structures.
- Temperature features: the temperature of the TIR anomaly is usually 2 ~6°C higher than that of outside or surrounding the TIR anomaly.
- Magnitude features: there are positive relations somewhat between the TIR anomaly energy (the anomaly area times the anomaly temperature) and the magnitude of future shock.

Anomalies Interpretation with RSRM-Model

The RSRM experimental results are applied to analyze satellite remote sensed TIR anomalies before several strong EQs in Asia. Referring to the seismogenic mechanisms, the satellite TIR anomalies are in good accordance with the detected IRR anomalies from rock fracturing and seismogenic experiments with fault system being simulated with disjointed faults and intersected faults.

Dongsha Ms5.9 EQ 1992 in Taiwan, China

Dongsha Ms5.9 EQ occurred in Taiwan on Sept 14, 1992. The NOAA satellite images show that there was TIR anomaly appeared along the regional faults and downfaulted basins before shock as in Figure 26 (Wu et al., 2006c). There was an isolated and spoon-shaped high temperature area to the southwest of Taiwan Island 25 days before shock (Aug 19, 1992). The head of 'spoon' locates in the downfaulted basin-IV and the handle of 'spoon' distributes along fault-6. Satellite TIR image on Aug 22 shows that the high temperature on 'spoon head' diminished, but the TIR anomaly on 'spoon handle' became wide. Later, the anomaly moves gradually to the epicenter. TIR image on Sept 9 shows that a large area of high temperature had appeared around the coming epicenter, and the maximum temperature (in brown color) appeared south-close to the coming epicenter. It indicated that the TIR anomaly was consistent with the regional tectonic structures (faults and basins) in spatial position and geometry.

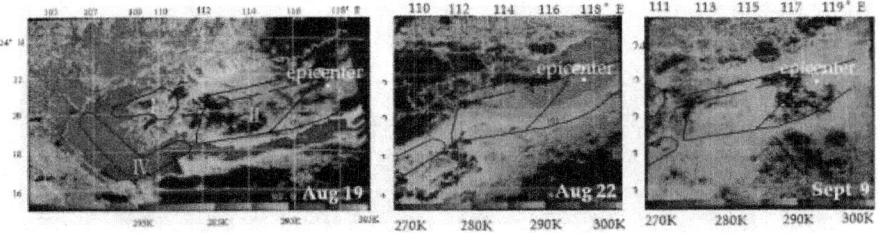

Figure 26. Satellite TIR anomaly images before Dongsha Ms5.9 EQ (Sept 14, 1992).

Zhangbei Ms6.2 EQ 1998 in China

Zhangbei Ms6.2 EQ occurred in China at 11:50 am, Jan 10, 1998. With the overlay of the active fault system investigated and deduced, in red lines, on night time NOAA satellite infrared images, as inFigure 27, it was discovered that (Wu et al., 2007b): 1) 18 days before shock, Dec 28, 1997, the TIR images on land and sea surface are basically normal, and the contours of TIR temperature field of Bohai Bay appeared to be along the coastline; 2) 5 days before shock, Jan 5, 1998, there occurred a positive TIR anomaly strip from Bohai Bay to Zhangbei passing through Beijing, its temperature was 3 C higher than that of outside, and the contours of TIR temperature field of Bohai Bay got offset the coastline; 3) 1 day before shock, Jan 8, 1998, the positive TIR anomaly strip had got wider and its temperature was 7 C higher than that of outside, and the contours of TIR temperature field of Bohai Bay got in accordance with the positive TIR strip; 4) 1 day after shock, Jan 11, 1998, the pattern of TIR images on the surface of land and sea turned to be normal again. There are several disjointed active faults along the positive TIR strip, including a possible uncovered great active deep fault going from Bohai Bay to Zhangbei. Besides, to the southwest of Zhangbei, there are two small active faults pointing to Zhangbei. If extend the three faults respectively towards Zhangbei, it could be found that Zhangbei is exactly the intersection point. Hence, the tectonic background around the epicenter is basically comprised of three intersected active faults as two groups, i.e., the primary fault is the independent one from Bohai Bay to Zhangbei, and the secondary two act as another group. The two groups of faults split basically the regional crust into three geo-blocks (geo-block A, B and C_1+C_2) as the RSRM experiment model in Figure 28b (rock block A, B and C), and the secondary two faults act as the two sides of an acute wedge-shaped geo-block (geo-block C_2) as the RSRM experiment model in Figure 28d (rock block C). The geometric features of the faults and the spatio-temporal features of TIR anomaly were similar. Therefore, the mechanical mechanism of Zhangbei EQ should be classified to be the failure of an intersected active fault system, which with an acute wedge-shaped geo-block being its secondary active object.

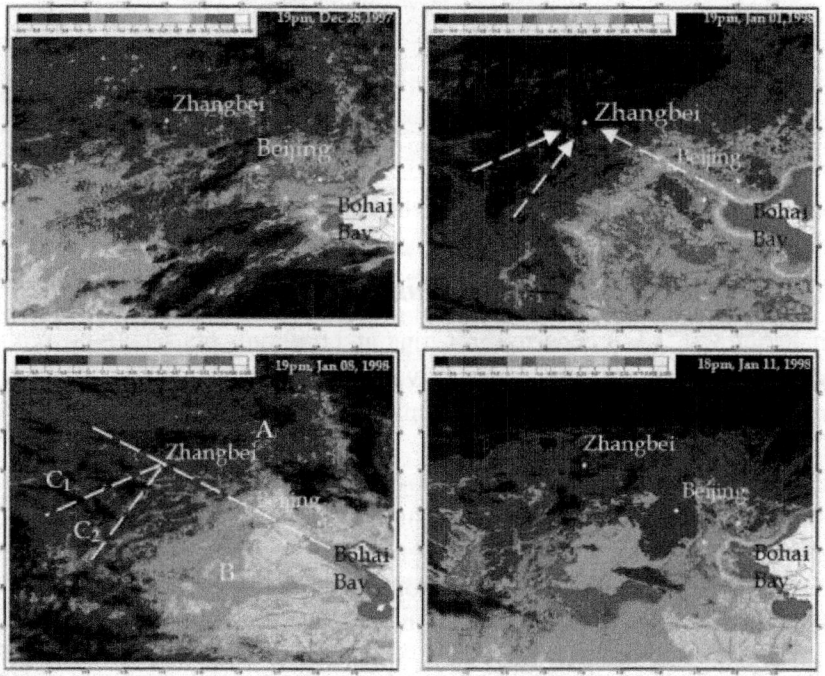

Figure 27. The TIR images overlaid with detailed active faults before and after Zhangbei Ms6.2 EQ (Jan 10, 1998).

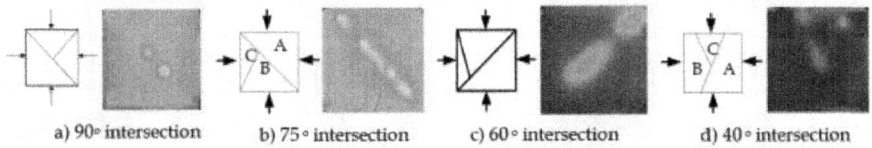

a) 90° intersection   b) 75° intersection   c) 60° intersection   d) 40° intersection

Figure 28. The IRR anomaly features of simulated tectonic activities due to bi-axially load (Wu et al., 2007b).

Izmit Ms 7.8 EQ 1999 In Turkey

Izmit Ms 7.8 EQ occurred in Turkey on Aug 17, 1999. In the epicenter zone, there were two en echelon tectonic faults, and the epicenter is 45km about to the west of the disjointed place of the two en echelon faults, as in Figure 29. With thermal image of Aug 1&2 being the reference, the differential thermal images from Aug 6 to Aug 26 were obtained, and it was revealed that there were NOAA satellite TIR anomaly at the disjointed zone of two en echelon faults since Aug 6, 11 days before shock (Tronin, 2000), as in Figure 29. The RSRM experiments on the simulation of tectonic EQ due to the fracturing of disjointed faults had revealed that there were IRR anomaly increment and concentrated deformation at the disjointed zone, as in Figure 30. Obviously, the spatial features of NOAA TIR anomaly before Izmit Ms 7.8 EQ were the same as that of IRR anomaly before the fracturing of disjointed faults.

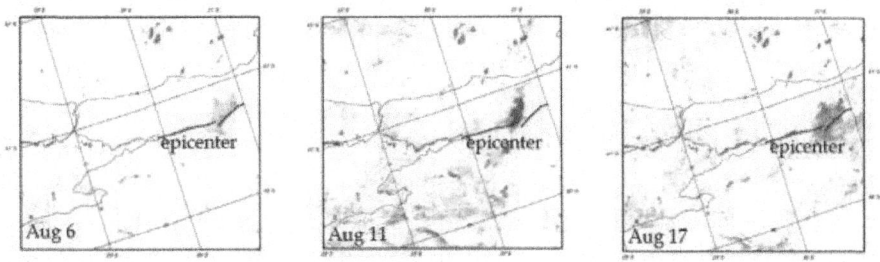

Figure 29. TIR anomalies 0~11 days before Izmit Ms 7.8 EQ (Aug 17, 1999).

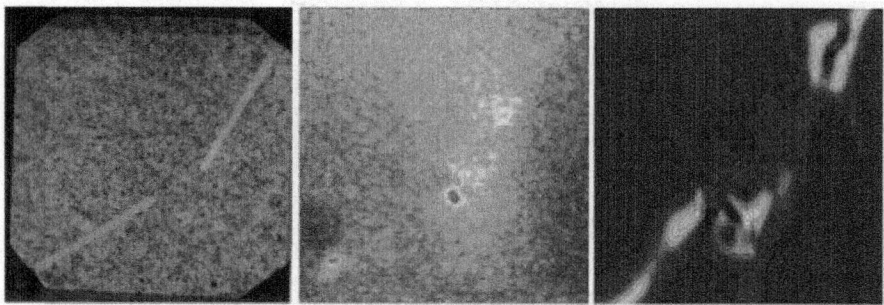

Figure 30. The IRR anomaly and deformation concentration of a bi-axially loaded granite sample with disjointed faults.

Hengchun Ms7.2 EQ 2006 in Taiwan, China

Hengchun Ms7.2 EQ occurred in Taiwan at 12:26pm, Dec 26, 2006. TIR images from stationary satellite FY-2 showed that there was TIR anomaly nearby the coming epicenter. The TIR anomaly appeared to the east of Philippines six days before shock, and moved gradually toward west to Philippines. Later, the anomaly changed its direction to the north, and the temperature increased 10°C about one day before shock. Figure 31 shows that the satellite TIR images around Hengchun on Dec 25, 2006 (Liu et al., 2007). A disjointed thermal strip (in orange color) appeared on the southwest of Taiwan at 1:00am, and developed gradually to be an X-shaped thermal anomaly zone at 10:00am. The epicenter located closely to the cross point of the X-shaped zones. The evolution process of satellite TIR images was very similar to that of X-shaped thermal IRR anomaly in RSRM experiments as inFigure 2.

Wenchuan Ms8.0 EQ 2008 in China

Wenchuan Ms8.0 EQ occurred in China at 14:28 am, May 12, 2008. The epicenter locates at the transferring zone of Tibetan Plateau to Sichuan Basin, which is only 92Km to the NW of Chengdu, the capital of Sichuan province. Analysis to FY-2C TIR images, as in Figure 32, it is discovered that there was an high temperature strip with length of 3 000 km appeared on April 23, 20 days before shock, which started from India Plate and developed to northeast along the east front of Tibetan and Loess Plateau (Wu et al., 2008). The cause might be the great accumulation of fictional sliding stress along the east foreland of Tibetan and Loess Plateau due to the subduction of India Plate to Euro-Asia Plate. The east foreland of Tibetan and Loess Plateau act as the fictional sliding plane between the west part of China (including Tibetan and Loess Plateau) and the east part of China (including the North, Middle and Southwest China).

The evolution of satellite TIR images was similar to that of detected strip-shaped IRR anomaly in the fictional sliding experiments as in Figure 8 and Figure 9, which shows the evolution of TIR anomaly along the sliding plane.

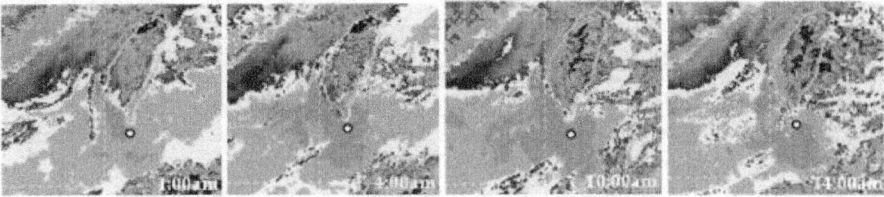

Figure 31. Satellite TIR images one day before Hengchun Ms7.2 EQ (Dec 25, 2006).

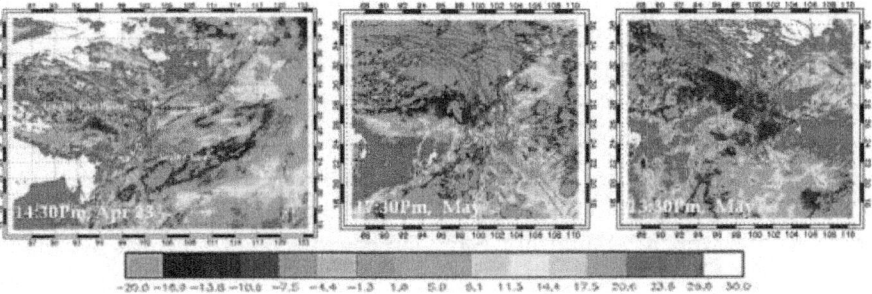

Figure 32. Satellite TIR images before Wenchuan Ms8.0 EQ (May 12, 2008).

FUTURE RESEARCHES

On RSRM Experiment

As a detectable remote sensing signal related with rock stress and physical temperature, IRR is a meaningful index for studying rock load, rock deformation, rock fracturing and rock hazard. The temporal evolution of surface IRR from loaded rock is the comprehensive effect of rock thermo-elastic acting, pore gas desorbing & escaping, fractures producing & extending, rock frictionating, heat transferring and environment radiation. The IRR image anomaly referring to the spatial-temporal evolution of IRR from loaded rock is an important precursor for rock fracturing, and will be meaningful for the predication of geo-hazards including tectonic EQ. For the practical application of RSRM, deeper research on IRR imaging detection quantitatively and specially on rock stress and rock hazard for experimental rock mechanics, rock engineering, tectonic activity and strong EQ is demanded.

The mechanism of experimental detected IRR anomaly can be theoretically interpreted by taking the load header, the rock sample and the environment to be a closed independent system in energy balance state. There are two of the main rock-physics mechanisms, respectively being thermo-mechanical coupling and frictional thermal due to tectonic stress, rock fracturing and fictional sliding, for the change IRR from loaded rock samples (Wu, et al., 2006c). Besides, positive hole (P-hole) activation due to piezoelectricity was suggested to be another mechanism of IRR from loaded quartz-bearing igneous rock (Freund, 2002), such as granite, basalt, diorite, and gabbro.

To search for scientific interpretation on the relations among satellite TIR anomaly, rock stress and experimentally detected IRR anomaly, the EM transferring process from underground rock body to satellite sensors, through lithosphere, earth surface coversphere (including soil layers, water bodies and vegetations), the atmosphere and lithosphere, should be systematically studied. Pulinets pointed out that the incubation of an EQ is to disturb the ionosphere (Pulinets, 1998), and it was suggested that lithosphere-atmosphere-ionosphere (LAI) coupling is the mechanism of satellite TIR anomaly before strong EQ (Molchanov et al., 2004). Nevertheless, the action of earth surface coversphere on the transferring and the magnifying of EM signals from underground loaded rock to atmosphere should not be ignored, even if its physical mechanism are not clear. For the scientific interpretation of satellite TIR anomaly before strong EQ, the lithosphere-coversphere-ionosphere (LCA) coupling is the key, while for the scientific interpretation of ionosphere anomaly, the lithosphere-coversphere-atmosphere-ionosphere (LCAI) coupling should be focused. However, present experiments on LAI, LCA and LCAI coupling are rather insufficient. Future experiments specially designed to uncover the mechanisms & laws and to construct the models & quantitative equations of LAI, LCA and LCAI couplings are expected.

On EQ Thermal Infrared Anomaly

Although there are uncertain influences from meteorological variation, satellite TIR anomaly has quite different identification features from that of unseismology-resulted TIR anomaly. Satellite TIR remote sensing is becoming a prospecting technique for monitoring tectonic activities and for predicting strong tectonic EQ, which provide a negativism to that EQ cannot be predicted. Anyway, the practical predication of EQ is not so easy. The regional tectonic background and the active fault system have extremely important affects on the incubation of EQ and the TIR anomaly. Especially, the intersected faults, compressively-sheared faults, and disjointed faults are to control the location and the routing of the spatial evolution of satellite TIR anomaly, and the brightest spot of TIR anomaly along the fault, or at the intersection point, or at disjointed zone of faults might foretell the possible epicenter (Wu et al., 2007b).

First of all, massive observation information including crust stress, land deformation, atmosphere components, underground water, surface and near-surface temperature, satellite remote sensing TIR anomaly and EM disturbance in ionosphere should be integrated together for data fusion and cross checking to analyze comprehensively the tectonic activity and rock fracturing process. A grid-based distributed database and analysis tools on TIR remote sensing images, with global and regional tectonic structures being its background, should be set up to assist the extraction of EQ TIR anomaly in different temporal and spatial scale. Besides, a quantitative model for tectonic activity analysis and for EQ magnitude predication based on TIR anomaly should be developed.

The GEOSS under construction is to provide an integral and integrated monitoring on earth environment, Geo-hazard and global disasters. A generalized remote sensing (GRS) based on the integration of space-based, aero-based, near-surface based, in-situ based and underground-based monitoring is forming in the world (Wu and Liu, 2007). The international broad and sincere cooperation, inside the framework of GEOSS without discipline exclusion and data privacy, between seismologists, remote sensing scientists, meteorologists, geophysical scientists, geochemist and spatial information scientists in good faith is expected. Besides, a powerful spatial information system, especially for EQ early warning and short-coming prediction based on GEOSS, should be developed. It should has powerful functions such as massive information classification, smart theme mapping, easy map-layer overlay, fast features extraction, effective data fusion, intelligent data mine, powerful knowledge discovery, and easy access and share.

A possible technical procedure based on GRS for satellite TIR anomaly monitoring, analyzing and early warning of tectonic EQ inside the framework of GEOSS might be that: 1) the seismological geology background being the foundation of satellite TIR anomaly analysis; 2) the long-term GPS continuous monitoring and D-InSAR measurement being the guidance of tectonic stress detection and active fault identification; 3) the underground water temperature, near surface air temperature, radon & green gas, structural cloud anomaly dairy monitoring being the forerunner for preliminary identification of coming EQ; 4) the anomaly analysis of satellite TIR, NCEP temperature and ionosphere disturbance being the dominant for early warning of temporal-spatial-magnitude parameters of EQ; and 5) the additional celestial stress on active faults being a special leading disturbance for possible tectonic EQ.

REFERENCES

1. K. Arun, C. Swapnamita, 2005 Thermal remote sensing technique in the study of pre-earthquake thermal anomalies. J. Ind. Geophys. Union, 9 3 197 207

2. B. Brady, G. Rowell, 1986 Laboratory investigation for the electrodynamics of rock fracture. Nature, 321 29 488 492

3. C. Cui, J. Zhang, Q. Xiao, 1999 Monitoring the thermal IR anomaly of Zhangbei earthquake precursor by satellite remote sensing technique, Proc. 20th Asia Remote Sensing Congress, 1179 1184 , Hongkong

4. M. Deng, J. Qian, J. Yin, et al. 2001 Research on the application of infrared remote sensing in the stability monitoring and unstability prediction of large concrete engineering. Chinese J Rock Mech. Engi., 20 2 147 50 .

5. F. Freund, 2002 Charge generation and propagation in igneous rocks. J. Geodynamics, 33(4-5): 543-570.

6. N. Geng, C. Cui, et. Deng, al, 1992 Remote sensing detection on rock fracturing experiment and the beginning of Remote Sensing Rock Mechanics. ACTA Seismologic SINICA, 14(supp) : 645 52 .

7. V. Gorny, A. Salman, A. Tronin, et al. 1998 The earth outgoing IR radiation as an indicator of seismic activity, Proc. Acad. Sci. USSR, 30 1 67 69 .

8. J. Hudson, J. Harrison, 1997 Engineering rock mechanics, Elsevier Science Inc., New York, 85 112

9. D. Liu, K. Peng, W. Liu, et al. 1999 Thermal omens before earthquake. Acta Seismologica Sinica, 12 6 710 715

10. S. Liu, L. Wu, J. Li, et al. 2007 Features and mechanism of the satellite thermal infrared anomaly before Henchun earthquake in Taiwan Region. Science & Technology Review, 25 6 32 37

11. S. Liu, L. Wu, Y. Wu, et al. 2002 Quantitative study on the thermal infrared radiation of dark mineral rock in condition of uni-axial loading. Chinese J. Rock Mech. & Engi., 21 11 1585 89

12. M. Luong, 1990 Infrared thermovision of damage processes in concrete and rock. Engi. Fracture Mechanics, 35 (1~3): 127-35.

13. G. Martelli, P. Smith, A. Woodward, 1989 Light, radiofrequency emission and ionization effects associated with rock fracture. Geophy. J. Int., 98 2 397 401 .

14. O. Molchanov, E. Fedorov, A. Schekotov, et al. 2004 Lithosphere-atmosphere-ionosphere coupling as governing mechanism for preseismic

short-term events in atmosphere and ionosphere. Natural Hazards and Earth System Sciences, 4 757 767

15. D. Mounatin, J. Webber, 1978 Stress pattern analysis by thermal emission (SPATE), Proc. Soc. Photo-Opt. Inst. Engi, 164:189.

16. A. Nicolas, J. Bouchez, J. Blaise, et al. 1977 Geological aspects of deformation in continental shear zones. Tectonophysics, 42 1 55 73

17. M. Ohtake, T. Matumoto, G. Latham, 1981 Evaluation of the forecast of the 1978 Oaxaca Southern Mexico earthquake based on a precursory seismic quiescence. In: Earthquake Prediction Maurice Ewing Series American Geophysics Union 4) , 53 62

18. D. Ouzounov, F. Freund, 2004 Mid-infrared emission prior to strong earthquakes analyzed by remote sensing data. Adv. Space Res. 33 268 273 .

19. S. Pulinets, 1998 Seismic activity as a source of the ionospheric variability. Adv. Space Res., 22 6 903 907

20. Z. Qiang, X. Xu, C. Dian, 1990 Thermal infrared anomaly-precursor of impeding earthquakes. Chinese Science Bulletin, 35 17 1324 1327

21. D. Renata, 1977 Electromagnetic phenomena associated with earthquakes. Geophsical survey, 3 2 157 174 .

22. R. Sibson, et al. 1980 Power dissipation and stress levels on faults in the upper crust. J. Geophysical research, 85(B11): 6239-6247.

23. W. Thomson, 1853 Trans. R. Soc. Edinburgh, 20 83 261 .

24. A. Tronin, 1996 Satellite thermal survey-a new tool for the studies of seismoactive regions. J Remote Sensing, 17 8 1439 1455

25. A. Tronin, 2000 http://www.iki.rssi.ru/earth/ppt/tronin.ppt

26. X. Wang, 2003 A research on the inorganic carbon dioxide from rock by thermal simulation experiments. Advance in Earth Science, 18 8 515 20

27. L. Wu, S. Liu, 2007a Generalized remote sensing for solid Earth hazards under condition of GEOSS. Science & Technology Review, 25 6 5 11

28. L. Wu, J. Wang, 1998 Infrared radiation features of coal and rocks under loading. Int. J. Rock Mech. & Min. Sci, 35 7 969 976

29. L. Wu, C. Cui, N. Geng, et al. 2000 Remote Sensing Rock Mechanics (RSRM) and associated experimental studies. Int J Rock Mech Min Sci, 37 6 879 88 .

30. L. Wu, J. Li, S. Liu, 2009 Infrared anomaly analysis based on reference fields for earthquake remote sensing. Seismology Geology, 2009(in press)

31. L. Wu, J. Li, X. Xu, et al. 2007b Theoretical analysis to impending tectonic earthquake warning based on satellite infrared anomaly, Proc. 2007 IEEE Int. Geosciences &Remote Sensing Symposium, 3723 3727

32. L. Wu, S. Liu, W. Shi, et al. 2003 Experimental study on infrared anomaly of tectonic earthquake, Proc. SPIE, Remote Sensing For Environmental Monitoring, GIS Applications & Geology III, 5239 376 387 .

33. L. Wu, S. Liu, Y. Wu, 2006c The experiment evidences for tectonic earthquake forecasting based on anomaly analysis on satellite infrared image, Proc. 2006 IEEE Int. Geosciences &Remote Sensing Symposium, 2158 2216

34. L. Wu, S. Liu, Y. Chen, et al. 2008 Satellite thermal infrared and cloud abnormities before Wenchuan earthquake. Science & Technology Review, 26 10 28 29

35. L. Wu, S. Liu, Y. Wu, et al. 2002 Changes in IR with rock deformation. Int. J. Rock Mech. & Min. Sci., 39 6 825 31

36. L. Wu, S. Liu, Y. Wu, et al. 2004a Remote Sensing Rock Mechanics (I) : laws of thermal infrared radiation from disjointed jointed faults fracturing and its meanings for tectonic earthquake omens. Chinese J Rock Mech. & Engi., 23 1 24 30

37. L. Wu, S. Liu, Y. Wu, et al. 2004b Remote Sensing Rock Mechanics (II) : laws of thermal infrared radiation from bi-sheared faults friction sliding and its meanings for tectonic earthquake omens. Chinese J Rock Mech. Engi., 23 2 192 198

38. L. Wu, S. Liu, Y. Wu, et al. 2004c Remote Sensing Rock Mechanics (IV): laws of thermal infrared radiation from compressively-sheared fracturing rock and its meanings for earthquake omens. Chinese J Rock Mech. & Engi., 23 4 539 544

39. L. Wu, S. Liu, Y. Wu, et al. 2006a Precursors for rock fracturing and failure-part I: IRR image abnormalities. Int J Rock Mech Mining Sci, 43 3 473 482

40. L. Wu, S. Liu, Y. Wu, et al. 2006b Precursors for rock fracturing and failure-part II: IRR T-curve abnormalities. Int J Rock Mech Mining Sci, 43 3 483 493

41. L. Wu, S. Liu, X. Xu, et al. 2004d Remote Sensing Rock Mechanics (III) : laws of thermal infrared radiation and acoustic emission from friction sliding intersected faults and its meanings for tectonic earthquake omens. Chinese J Rock Mech. & Engi., 23 3 401 407

42. X. Xu, X. Xu, Y. Wang, 2000 Satellite infrared anomaly before Nantou

Ms=7.6 earthquake in Taiwan, China. ACTA Seismologica SINICA, 22 6 656 659

43. I. Yamada, K. Masuda, H. Murakami, 1989 Electromagnetic and acoustic emission associated with rock fracture. Phys. Earth Planet. Inter., 57(1-2):157-168.

44. Y. Zhang, W. Shen, 2002 Satellite thermal infrared anomaly before the Xinjiang Qianhai boder Ms8.1 earthquake. Northwestern Seismological J., 34 1 1 4

Chapter 3

ROUGHNESS RESEARCH OF CENTER PROFILE CURVE ON ROCK FRACTURE SURFACE BASED ON STATISTICAL METHOD

Xuezai Pan[1,3], Zhigang Feng[2,3], Guoxing Dai[3], and Hongguang Liu[4]

[1]School of Mathematics, Nanjing Normal University, Taizhou College, Taizhou, China

[2]State Key Laboratory of Coal Resources and Safe Mining, China University of Mining and Technology, Beijing, China

[3]Faculty of Science, Jiangsu University, Zhenjiang, China

[4]Faculty of Civil Engineering and Mechanics, Jiangsu University, Zhenjiang, China

ABSTRACT

In order to research roughness of rock fracture surfaces whether to depend on scale effect, Brazil discs were fractured under tensile and compression stresses in Brazil split test with MTS (Mechanics Test Systems) and a laser profilometer was used to scan rock fracture surfaces and coordinates datum of central profile were acquired. A figure of the central profile was plotted through the coordinates datum. A certain line segment length is regarded as a step length, which is called scale and the scale length is taken to connect pairs of closer peak points on the profile curve. The directional distribution of every scale's normal vector is analyzed by statistics and normal hypothesis test. Finally, some statistics of sample degrees datum are compared with other ones and reach a conclusion that roughness of center profile curve depends on scale effect. The distribution of degrees more and more approximates normal distribution along with increase of scale.

INTRODUCTION

Deformity and fracture of rock are involved in process of moving in earth crust, for example, earthquake, slide downhill, mud-stone flow and so on. In addition, rock fracture usually happens in rock project, for instance, explosion of rock, the project of tunnel, mining engineering etc. Studying rock fracture surfaces through morphology has been recognized by professionals since twenty-one century, because morphology of rock fracture surfaces implicates abundant information of rock fracture mechanics. Experts have discovered that rock fracture surface has characterization of roughness, irregularity and complexity. They attempt to depict the relation between complex morphology of rock fracture surfaces and roughness by all kinds of ways. For example, In order to assess the current state of rock masses and to predict the stability of jointed rock structures, the roughness of rock fracture surfaces has been studied to a higher level. Based on systematic experiments, Barton and Choubey in 1977 proposed a conceptual model to quantify the roughness of rock fracture surface [1]. They classified the roughness into ten categories and the Joint Roughness Coefficients (JRC) ranged from 0 to 20 [2-4]. Some investigators use fractal geometry and multifractal which have been developed since 1970's to describe rock fracture mechanism and have tried to establish the relationship between the fractal dimension and various mechanical parameters of rock fracture surfaces [5-8]. Some experts have indicated that the structural anisotropy of fracture surfaces in rocks greatly influences the mechanical behavior of rock joints under loading [9-11]. The following statement will study roughness of center profile curve on rock fracture surfaces from statistical view. Finally, three prospects will be put forward in the end of the paper.

EXPERIMENTAL METHOD AND ANALYSIS

Experimental Method

Firstly, a sort of special granites which were taken from Gansu province North Mountain in China were used to experimental material, because the compactness of the rock material is relatively homogeneous. The granite material was made of the cylinder-shaped sample with rock drilling machine, and then the cylinder-shaped sample was cut into three Brazil discs samples with cutting off machine and buffing machine. The diameter and the height of the discs are equal to 112 mm and 28 mm respectively. Secondly, the rock discs were fractured under tensile and compression stresses in Brazil split test with MTS (Mechanics Test Systems). Loading speed was perminute 0.01 mm. When loading strength approximatively reaches 48 kN, the discs were fractured along vertical direction (refer with: **Figure 1**). Finally, according to rock mechanics principle, in indirect tensile stresses process of rock, the rock stress of the edge of disc is relatively centralized, so the edge of the disc was easily broken and a little stone chips fell. Whereas inner stress of the rock disc is relatively balanced [12-13], the inner of fracturerock has no stone chips fallen. So, 11 mm was removed from two ends of the rectangular fracture surfaces respectively. The length of center part of fracture surface is equal to 90 mm (refer with: **Figure 2**). The length of 90 mm is supposed to x axis direction.

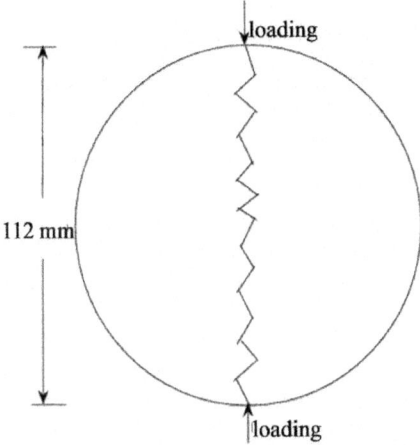

Figure 1. Indirect tensile diagrammatic sketch.

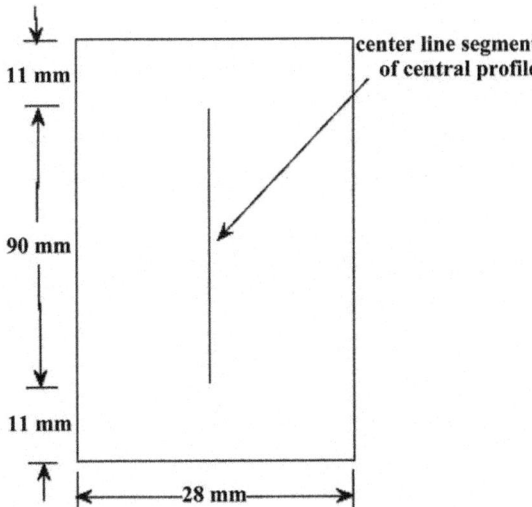

Figure 2. Center profile acquired datum.

The center part of fracture surface was scanned by high-accuracy rock laser profilometer along x axis according to the way that interval of x axis is equal to 0.1 mm to acquired three dimension coordinates (x, y, z) of lattice. Total 901 rope line segments were scanned through the above method, because the length of center part of fracture surface is 90 mm. Length of every line segment scanned is 28 mm, since the width of center part of rock fracture surface is equal to 28 mm (refer with: **Figure 2**).

Center Profile Curve Analysis on Rock Fracture Surface

The coordinates datum of the center segment of central profile curve on rock rectangle fracture surface (refer with: **Figure 2**) was extracted by computer procedure, and then the approximative two dimension curve figure of the profile was acquired by linear interpolating method (refer with: **Figure 3**). A unit length of step length is supposed to 0.1 mm. A step length was taken to connect pairs of points on the center profile [14-16]. Normal vectors' directional distribution of corresponding step length was considered and every angle of each normal vector departing from straight up vector is measured (refer with: **Figure 4**).The motivation for this approach is straightforward: for a perfectly straight profile the normal vectors will all be parallel and hence display zero dispersion, whereas the dispersion will increase as the profile departs from straight (i.e. becomes rougher). Suppose angle of straight up vector is zero degree. The degree of the angle in the direction skewing to left is negative, whereas that skewing to right is positive. Thus, these degree datum of angles can be obtained with computer procedure.

STATISTICAL METHOD

From statistical knowledge [17-20], mean and median of sample (refer with: Equation (1)) reflect concentrated tendency of sample datum.

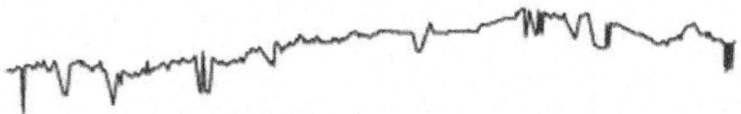

Figure 3. Linear interpolating figure of center profile.

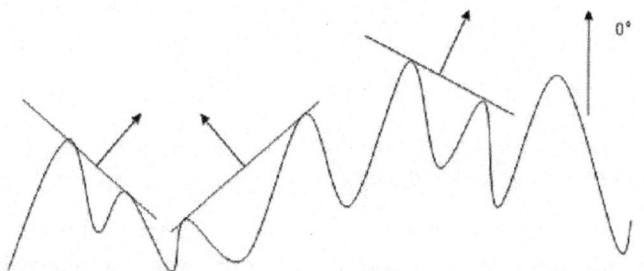

Figure 4. Orientation of normal vector to a step length connecting two points on a profile curve.

Range (refer with: Equation (2)), variance and sample standard deviation (refer with: Equation (3)) indicate the extent that sample datum depart from sample mean. Skewness (refer with: Equation (4)) and kurtosis (refer with: Equation (5)) are such statistic describing the shape of sample datum. Skewness reflects dispersive symmetrical characterization of sample datum. Finally, kurtosis indicates the situation that sample datum deviate normal distribution.

$$\bar{X} = \frac{1}{n}\sum_{i=1}^{n} X_i \tag{1}$$

$$R = \max(X_i) - \min(X_i) \tag{2}$$

$$s = \sqrt{\frac{1}{n-1}\sum_{i=1}^{n}(X_i - \bar{X})^2} \tag{3}$$

$$g_1 = \frac{1}{s^3}\sum_{i=1}^{n}(X_i - \bar{X})^3 \tag{4}$$

$$g_2 = \frac{1}{s^4}\sum_{i=1}^{n}(X_i - \bar{X})^4 \tag{5}$$

where X_i denotes samples.

When $g_1 > 0$, the form is called right deviation, which illustrates the right datum of mean are more dispersive than those of the left datum; As $g_1 < 0$, the result is named left deviation, which illustrates the situation is opposite to that of right deviation. As g_1 approach zero, which is called impartiality, So, the distribution is regarded as symmetry. On the other hand, kurtosis of normal distribution is equal to 3. As $g_2 > 3$, there are a lot of datum departing from mean, whose shape of distributive curve is flatter than that of normal distribution accordingly; On the contrary, when $g_2 < 3$, the case is inverse to that of $g_2 > 3$. So, statistical method can be used to characterize roughness of profile curve subjected to fracture surfaces. The following discussion is concrete operation.

Suppose the step length is equal to 0.4 mm, 0.3 mm, 0.2 mm and 0.1 mm respectively, then the datum of angle variation are acquired by computer procedure corresponding to various step length. Furthermore, under the same scale, sample mean, median, range, variance, standard deviation, coefficient of skewness and kurtosis are computed respectively and frequency histogram [21-25] is drawn with computer program, which describes the distribution of orientation of normal vectors from the center profile curve. Frequency histogram under a step length is compared with that of other ones, which can show distributional differences each other. On the other hand, hypothesis test method is used to test whether the distribution of normal vectors obeys normal distribution or not and distribution function plots were drawn with computer program. For the degree datum input into computer, Jarque-Bera test is used to test these degree datum whether to obey normal distribution. Significance level α is supposed to 0.05. P is a probability value accepting original hypothesis. JBSTAT is test statistics value. CV is a threshold which can judge whether to refuse original hypothesis and H is test result. If H = 0, the distribution of the degree datum can be considered normal distribution; If H = 1, the distribution of the degree datum doesn't obey normal distribution. If $P < \alpha$, original hypothesis that the datum belong to normal distribution can be denied; If JBSTAT > CV, normal distributional original hypothesis can be negated. Every statistic in following tables is a mean value of corresponding statistic of three center profile curves datum under the same step length (i.e. the same scale), because there are three Brazil discs samples. The differences among the same statistic are compared within four tables under different scales.

1) If the step length is equal to 0.4 mm, the following Tables 1 and 2 indicate a statistical result.

In **Table 1**, unit of mean, median, range and standard deviation is degree, whereas other statistics have no unit, because they are only coefficients (below affinity).

Variables in **Table 2** have no unit (below affinity). From coefficient of skewness −0.0013 ($\approx$0), dispersive extent of datum with left side and right one deviating mean is almost comparative. Distribution of angles' degrees approximatively summits to normal distribution from kurtosis coefficient 3.1511 ($\approx$3) and its frequency histogram is referred with **Figure 5**. From hypothesis test view, where H = 0, $P < \alpha$ and JBSTAT < CV, the normal distributional original hypothesis can be accepted. From normal probability plot shown in **Figure 6**, the absolute major points gather on the red straight line, which illustrates the normal distributional suppose can be accepted.

In **Figure 5**, i denotes positive integer and $1 \leq i \leq 12$, because frequency histogram consists of twelve columns.

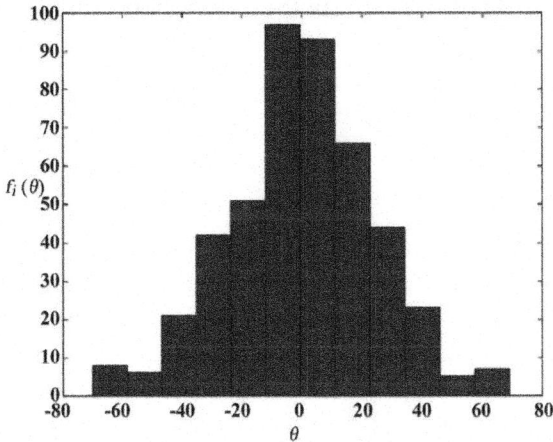

Figure 5. The histogram with the step length 0.4 mm (θ: degree; $f_i(\theta)$: frequency).

Table 1. Statistics of sample datum with the step length 0.4 mm.

Mean	Median	Range	Variance	Standard deviation	Skewness	Kurtosis
−0.1333	0.0000	149.0381	603.3292	24.5628	−0.0013	3.1511

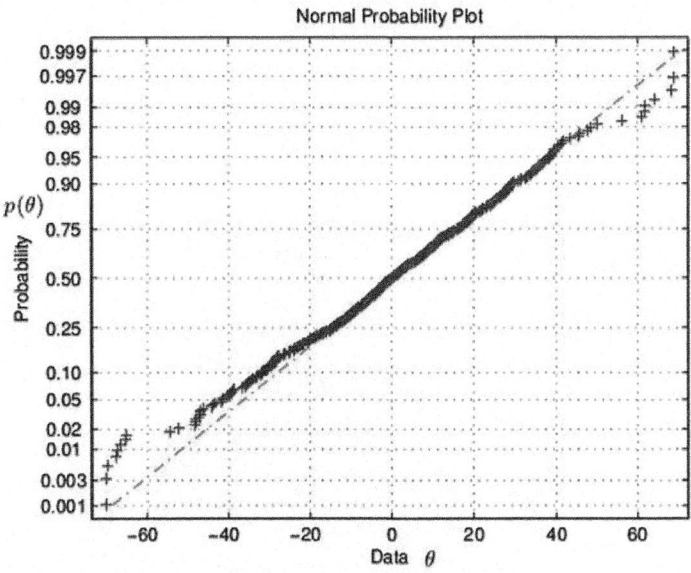

Figure 6. Normal probability plot with the step length 0.4 mm (θ: degree; $p(\theta)$: probability).

Table 2. Hypothesis test value with the step length 0.4 mm.

H	P	JBSTAT	CV
0	0.2314	2.9269	5.9915

2) If the step length is equal to 0.3 mm, the statistical datum result is shown in the Tables 3 and 4.

The extent of departing from sample mean increases; The skewness coefficient increases and is more than 0, which illustrates the right datum of mean is more dispersive than that of the left, but the dispersive extent is faint; Kurtosis coefficient is equal to 2.8296 ($\approx$3), which illustrates distribution of angle datum approximate normal distribution. The frequency histogram reflects that the distribution of angle datum close to normal distribution (shown in **Figure 7**). From normal hypothesis test view, H = 0, P = 0.6492 and JBSTAT < CV indicate normal distributional hypothesis can be accepted with 64.92% probability. From normal probability plot shown in Figure 8, the absolute major points gather on the red straight line, which illustrates the normal distributional suppose can be accepted.

3) If the step length is equal to 0.2 mm, the statistical result is shown in the next Tables 5 and 6.

The sample mean and median increase; Variance and standard deviation increase furthermore; Range hardly change; Coefficient of skewness reduces, however the decrement is very little; coefficient of kurtosis decreases furthermore and reaches 1.9127, which illustrates the distribution of angle datum continues to deviate from normal distribution.

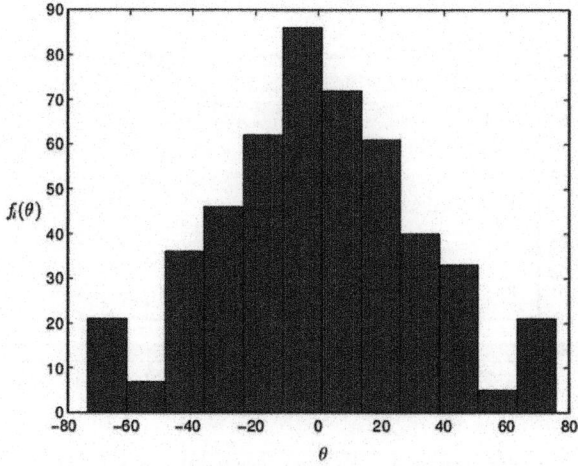

Figure 7. The histogram with the step length 0.3 mm (θ: degree; $f_i(\theta)$: frequency).

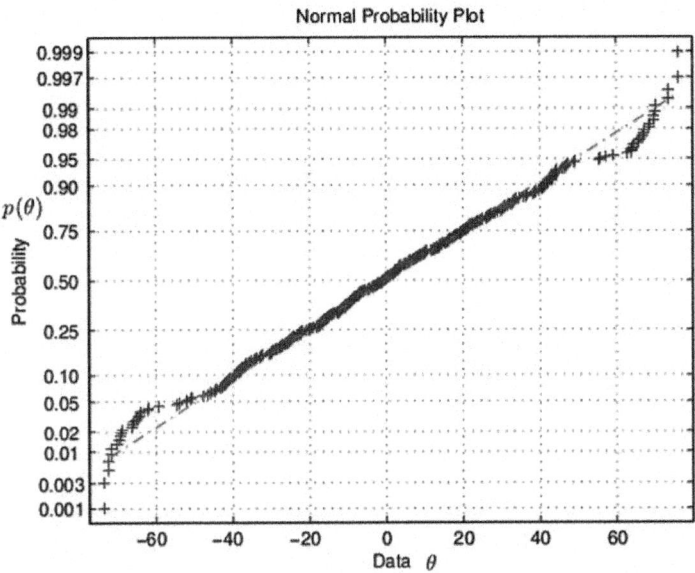

Figure 8. Normal probability plot with the step length 0.3 mm (θ: degree; p(θ): probability).

Frequency histogram is shown in **Figure 9**. From normal hypothesis test view, H = 1, the value of P and JBSTAT > CV indicate normal distributional hypothesis can be negated. Normal probability plot is referred with **Figure 10**.

4) If the step length is equal to 0.1 mm, the following Tables 7 and 8 indicate according statistical result.

Table 3. Statistics of sample datum with the step length 0.3 mm.

Mean	Median	Range	Variance	Standard deviation	Skewness	Kurtosis
−0.2117	−0.8593	149.5869	990.3574	31.4699	0.0488	2.8296

Table 4. Hypothesis test value with the step length 0.3 mm.

H	P	JBSTAT	CV
0	0.6492	0.8640	5.9915

Table 5. Statistics of sample datum with the step length 0.2 mm.

Mean	Median	Range	Variance	Standard deviation	Skewness	Kurtosis
0.2166	0.0000	149.5887	1588.7498	39.8591	0.0081	1.9127

Table 6. Hypothesis test value with the step length 0.2 mm.

H	P	JBSTAT	CV
1	2.5850e−011	48.7574	5.9915

Table 7. Statistics of sample datum with the step length 0.1 mm.

Mean	Median	Range	Variance	Standard deviation	Skewness	Kurtosis
0.1318	0.2865	149.5898	1783.2761	42.2289	0.0072	1.6600

Table 8. Hypothesis test value with the step length 0.1 mm.

H	P	JBSTAT	CV
1	0	112.9056	5.9915

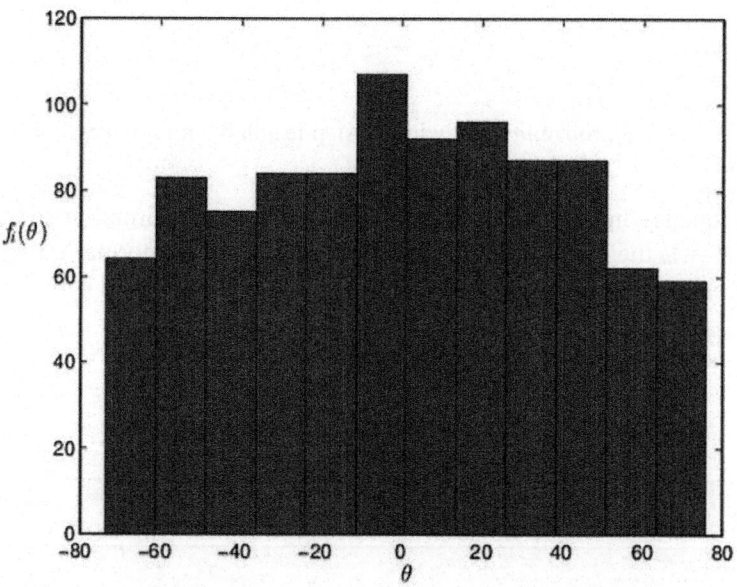

Figure 9. The histogram with the step length 0.2 mm (θ: degree; $f_i(\theta)$: frequency).

The value of sample mean and median has a weak variation; Range still change rarely; Variance and standard deviation increase greatly; Coefficient of skewness continues to decrease, but it still fluctuates near 0, which still describes that two sides' datum of mean have the same dispersion characterization; Coefficient of kurtosis descends further and attains 1.6600, which indicates that the distribution of normal vector deviates from normal distribution; Frequency histogram is shown in **Figure 11**.

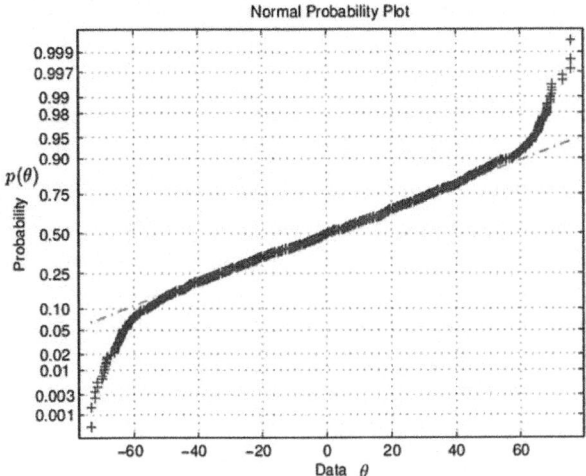

Figure 10. Normal probability plot with the step length 0.2 mm. (θ: degree; p(θ): probability).

From normal hypothesis test view, H = 1, the value of P = 0 and JBSTAT ≫ CV indicate normal distributional hypothesis can be negated completely. Normal probability plot is referred with **Figure 12**.The absolute major points deviates from the red straight line.

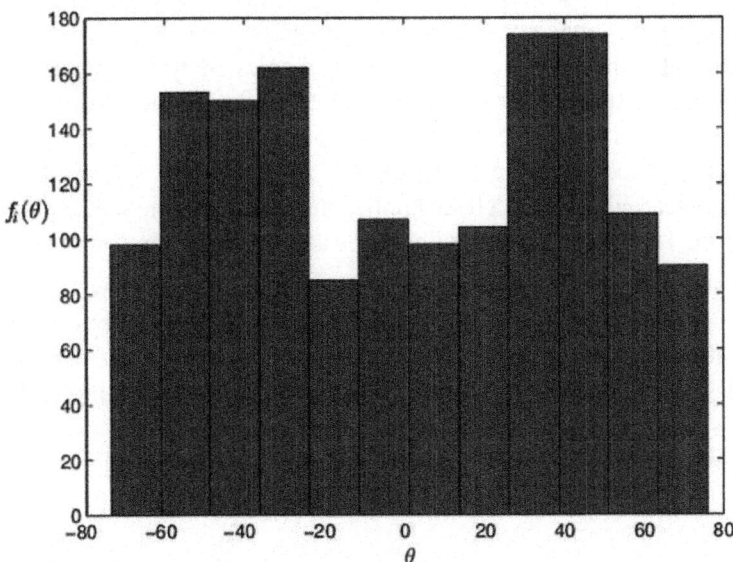

Figure 11. The histogram with the step length 0.1 mm (θ: degree; $f_i(θ)$: frequency).

In conclusion, the directional distribution of normal vectors is associated with the step length, which depends on scale effect. The smaller the kurtosis is, the smaller the scale is, which illustrates the distribution of angle datum deviates from normal distribution continuously. The case that range of sample datum is close to 149 is disco vered, which indicates that variable range of angle degrees is definite. And the dispersive extent between the left datum of sample mean and those of right is comparative, because skewness coefficient is almost equal to 0. In summary, the roughness of center profile curve could be supposed to standard state, when the distribution of angle datum belongs to standard normal distribution. Furthermore, the rougher center profile curve is, the smaller the step length is.

CONCLUSION AND PROSPECT

In generally, mean, median, range and skewness coefficient almost have no scale effect, however, variance and standard deviation will increase with decreasing of scale. Kurtosis coefficient will descent along with diminishing of scale. Accordingly, from normal hypothesis test view, H, P, JBSTAT have scale effect too. As a whole, H leaps from 0 to 1 along with decreasing of scale. P will decrease while scale drops. JBSTAT will raise as scale falls. That is to say, the rougher center profile curve is, the smaller the step length is. The ultimate purpose of researching morphology of rock fracture surfaces is that the information in process of rock fracture is acquired through methods of mathematical analysis. Successively, components of rock structure and limitation are discovered and further mechanics of rock fracture is posted. However, structure of rock and properties of mechanics exhibit nonlinear characterization. Rock fracture surface possesses fairly irregular and stochastic properties, hence the topic is developed so slowly that the results acquired through researching haven't been applied in forecasting and instructing practise of project. Therefore, the following three aspects of research will be evolved.

Firstly, advantages of approaches applied by experts now continue to be developed and perfected, and insufficiencies will be overcome extensively, which attempt to build the relationship between rock mechanics and topograph of rock fracture surfaces. Secondly, new approaches of studying rock fracture surfaces through morphology will be sought, which try to discover the mechanics of rock fracture. Finally, existing experimental results will be transformed into theoretical basis of guiding project practice possibly.

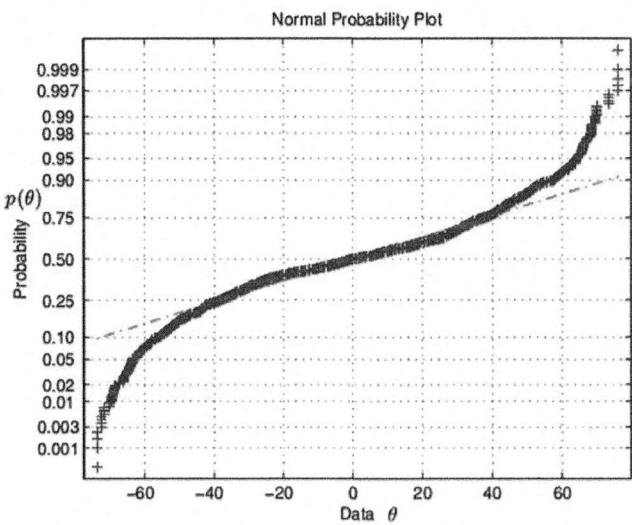

Figure 12. Normal probability plot with the step length 0.1 mm (θ: degree; $p(\theta)$: probability).

ACKNOWLEDGEMENTS

- Supported by the National Natural Science Foundation of China 51079064.

- Supported by State Key Laboratory of Coal Resources and Safe Mining, China University of Mining and Technology SKLCRSM10KFA02.

- Supported by Nanjing Normal University Taizhou College Q201234.

REFERENCES

1. N. Barton and V. Choubey, "The Shear Strength of Rock Jounts in Theory and Practice," Rock Mechanics, Vol. 1977, Vol. 10, No. 2, pp. 1-45. doi:10.1007/BF01261801

2. N. Barton, "Review of a New Shear Strength Criterion for Rock Joints," Engineering Geology, Vol. 7, No. 4, 1973, pp. 287-332. doi:10.1016/0013-7952(73)90013-6

3. R. Tse and D. M. Cruden, "Estimating Joint Roughness Coefficients," International Journal of Rock Mechanics and Mining Sciences & Geomechanics Abstracts, Vol. 16, No. 5, 1979, pp. 303-307. doi:10.1016/0148-9062(79)90241-9

4. N. H. Maerz, J. A. Franklin and C. P. Bennett, "Joint Roughness Measurement Using Shadow Profilometry," International Journal of

Rock Mechanics and Mining Sciences & Geomechanics Abstracts, Vol. 27, No. 5, 1990, pp. 329-343. doi:10.1016/0148-9062(90)92708-M

5.   K. Falconer and W. G. Yang, "Fractal Geometry Mathematical Foundations and Applications," 2nd Edition, Posts & Telecom Press, Beijing, 2007.

6.   Y. H. Zhang, H. W. Zhou and H. P. Xie, "Improved Cubic Covering Method for Fractal Dimensions of a Fracture Surface of Rock," Chinese Journal of Rock Mechanics and Engineering, Vol. 24, No. 17, 2005, pp. 3192- 3196.

7.   H. P. Xie, H. Q. Sun, Y. Ju, et al., "Study on Generation of Rock Surfaces by Using Fractal Interpolation," International Journal of Solid and Structure, Vol. 38, No. 32-33, 2001, pp. 5765-5787. doi:10.1016/S0020-7683(00)00390-5

8.   H. P. Xie, J.-A. Wang and M. A. Kwa Niewski, "Multifractal Characterization of Rock Fracture Surfaces," International Journal of Rock Mechanics Mining Sciences, Vol. 36, No. 1, 1999, pp. 19-27. doi:10.1016/S0148-9062(98)00172-7

9.   S. C. Bandis, A. C. Lumsden and N. R. Barton, "Fundamentals of Rock Joint Deformation," International Journal of Rock Mechanics and Mining Sciences & Geomechanics Abstracts, Vol. 20, No. 6, 1983, pp. 249-268. doi:10.1016/0148-9062(83)90595-8

10.   P. H. S. W. Kulatilake, G. Shou, T. H. Huang and R. M. Morgan, "New Peak Shear Strength Criteria for Anisotropic Rock Joints," International Journal of Rock Mechanics and Mining Sciences & Geomechanics Abstracts, Vol. 32, No. 7, 1995, pp. 673-697.doi:10.1016/0148-9062(95)00022-9

11.   H. W. Zhou and H. Xie, "Anisotropic Characterization of Rock Fracture Surfaces Subjected to Profile Analysis," Physics Letters A, Vol. 325, No. 5-6, 2004, pp. 355-362.doi:10.1016/j.physleta.2004.04.006

12.   W. Gao, "Mechanics of Rock," Peking University Press, Beijing, 2010.

13.   Z. L. Fu, F. K. Xiao, Y. X. Liu and S. J. Chen, "Experiment Course on Rock Mechanics," Chemistry Engineering Press, Beijing, 2010.

14.   V. Rasouli and J. P. Harrison, "Assessment of Rock Fracture Surface Roughness Using Riemannian Statistics of Linear Profiles," International Journal of Rock Mechanics Mining Sciences, Vol. 47, No. 6, 2010, pp. 940-948. doi:10.1016/j.ijrmms.2010.05.013

15.   V. Rasouli and J. P. Harrison, "Scale Effect, Anisotropy and Directionality of Discontinuity Surface Roughness," Proceedings of the EUROCK

Symposium, Aachen, 27-31 March 2000, pp. 751-756.

16. T. H. Wu and E. M. Ali, "Statistical Representation of Joint Roughness," International Journal of Rock Mechanics and Mining Science & Geomechanics Abstracts, Vol. 15, No. 5, 1978, pp. 259-262. doi:10.1016/0148-9062(78)90958-0

17. Z. S. Wei, "Probability and Statistics Tutorial," Higher Education Press, Beijing, 1983.

18. X. R. Chen, "Higher Statistics," China University of Science and Technology, Hefei, 1999

19. G. X. Liu, Z. F. He and J. L. Yang, "Probability and Statistics," Gansu Education, Lanzhou, 2002.

20. X. S. Liu, "Probability and Statistics," Sichuan University Press, Chengdu, 2009.

21. L. B. Wu and B. N. Li, "Mathematical Experiment and Modeling," Defense Industry Press, Changsha, 2007.

22. J. Zhao and Q. Dan, "Mathematical Modeling and Mathematical Experiment," Higher Education Press, Beijing, 2000.

23. H. G. Zhang, "Practical Tutorial of MATLAB/SIMULINK," Posts & Telecom Press, Beijing, 2009.

24. D. X. Zhang and L. S. Zhao, "Teaching of Language Procedure Design of MATLAB," China Railway Press, Beijing, 2010.

25. P. Wu, "Technology and Application of Effective Procedure of MATLAB: Analysis of 25 Cases," Beihang University Press, Beijing, 2010.

Chapter 4

CORRELATION OF SEISMIC P-WAVE VELOCITIES WITH ENGINEERING PARAMETERS (N VALUE AND ROCK QUALITY) FOR TROPICAL ENVIRONMENTAL STUDY

Andy A. Bery, Rosli Saad

Geophysics Section, School of Physics, Universiti Sains Malaysia, Penang, Malaysia

ABSTRACT

The physical parameters of the subsurface from the environmental site investigation are important for geoscientists and engineers to understand and very low cost-effective method, especially when combined with geophysical (seismic) and geotechnical (borehole) surveys. These parameters can be estimated from other obtained parameters. In this study, P-wave velocities of materials (soils and rocks) are studied both in the laboratory and field measurement. The obtained P-wave velocities are then compared with the engineering parameters such N values, rock quality, friction angle, relative density, velocity index, density and penetration strength from boreholes. The empirical correlations were also found in this study for selected parameters. The estimation of engineering parameters from P-wave seismic velocity values is applicable for tropical environmental study. It is found that, the ratio (V_{FIELD}/V_{LAB}) when squared, was numerically close to the value of percentage RQD. We found that the empirical correlation for tropical environmental study is $V_p = 23.605(N) - 160.43$ and the regression found is 0.9315 (93.15%). Meanwhile, the empirical correlation between P-wave velocities and RQD values is found as $V_p = 21.951(RQD) + 0.1368$ and the regression found is 0.8377 (83.77%). The correlation between apparent P-wave velocities with penetration strength for both study sites are found as $V_{P(APP)} = 292.59(Ps) + 474.69$ and the regression coefficient is found as 0.9756. Thus, this study helps for the estimation and prediction the properties of the subsurface material (soils and rocks) especially in reducing the cost of investigation and increase the understanding of the Earth's subsurface characterizations physical parameters.

INTRODUCTION

In environmental study, there are many physical parameters can take into our consideration before making important decision for civil construction. These physical parameters play important role in indicating their behaviour which is due to time and condition changes. In this study, the data for the seismic refraction (tomography) method is correlated with the borehole data collected from the study site. The correlation found in this study could help in determine the signature of engineering parameters from infield data. Effort to relate rock quality and seismic velocity have been made at intervals, during the development and integration of rock engineering and engineering geology.

The in-situ behaviour of soils is complex because it is heavily dependent upon numerous factors. To acquire appropriate understanding, it is necessary to analyze them not only through geophysics and geotechnical engineering skills but also through other associated disciplines like geology, geomorphology, climatology and other earth and atmosphere related sciences [1].

Seismic refraction investigates the subsurface by generating arrival time and offset distance information to determine the path and velocity of the elastic disturbance in the ground. The disturbance is created by shot, hammer, weight drop, or some other comparable method for putting impulsive energy into the ground. Detectors lie out at regular intervals; measure the first arrival of the energy and its time. The data are plotted in time—distance graphs from which the velocities of the different layers and their depths can be calculated.

We believe that a deeper understanding of the seismic process can contribute to improved interpretations. Much information on later seismic arrivals is now attainable by use of the engineering seismographs which display the complete seismic waveform. The interpreter will be unable to understand these arrivals without knowing something about possible seismic wave types, seismic wave parts and expected travel-time patterns. Seismic images have become more accurate with the development of more sophisticated velocity models, which contain information about the speed with which the seismic waves travel through the rock layers. This information is critical for unravelling the geologic secrets hidden below [1].

In the last decade for instance, research studies on soil (sand and fines) characterizations. The role of fines in liquefaction mechanism is not fully understood yet [2]. These fines may affect the compression characteristics of coarse grained soils as well. In some models proposed for compression behaviour of cohesionless soils such as those by [3]; effects of initial void ratio, relative density, particle shape, mineralogy, structure and applied stress conditions were mentioned. These factors were also prominent in the experimental researches related to the compression of sands [4]. Study in basic engineering properties such as the grain size distributions, hardness, strength, durability and shear strength parameters (cohesion, C' and friction angle, φ') is important to understand the behaviour for older alluvium and avoid the inherence problems [5]. Many previous researchers [6,7] studied the changes of engineering properties for igneous and sedimentary rocks but very minimal works has been carried out for older alluvium.

The main objective of this study is to determine the correlation in between seismic velocity values with engineering parameters such as N value, rock quality, friction angle, relative density, strength (force), consistency and velocity index. Beside than that, the correlation found also extent for good estimation which is important in engineering perspective especially for tropical region country.

GEOGRAPHY AND GEOLOGY OF STUDY AREA

Our field investigation areas are located at south of Penang Island and Sarawak Malaysia. The major portion of Penang Island is underlain by igneous rocks (**Figure 1**). All igneous rocks are granites in terms of Streckeisen classification [8]. These granites can be classified on the basis of proportions of alkali feldspar to total feldspars. On this basis granites of Penang Island are further divided into two main groups: the North Penang Pluton approximately north of latitude 5°23› and the South Penang Pluton. In the northern part of the island, the alkali feldspars that generally do not exhibit distinct crosshatched twining are orthoclase to intermediate microcline in composition. In the southern region, they generally exhibit well-developed cross-hatched twining and are believed to be microcline. The North Penang Pluton has been divided into Feringgi Granite, Tanjung Bungah Granite and Muka Head micro granite. The South Penang Pluton has been divided into Batu Maung Granite and Sungai Ara Granite [8].

Rock exposures around the Miri Town, Sarawak, and belonging to the Miri Formation (middle Miocene strata); represent the uplifted part of the subsurface sedimentary strata [9]. The Middle Miocene sandstones of the Miri Foundation are located at north-eastern Sarawak, east Malaysia (**Figure 2**). The Miri Formation comprises sandstone, sand and clays in varying proportions and thickness.

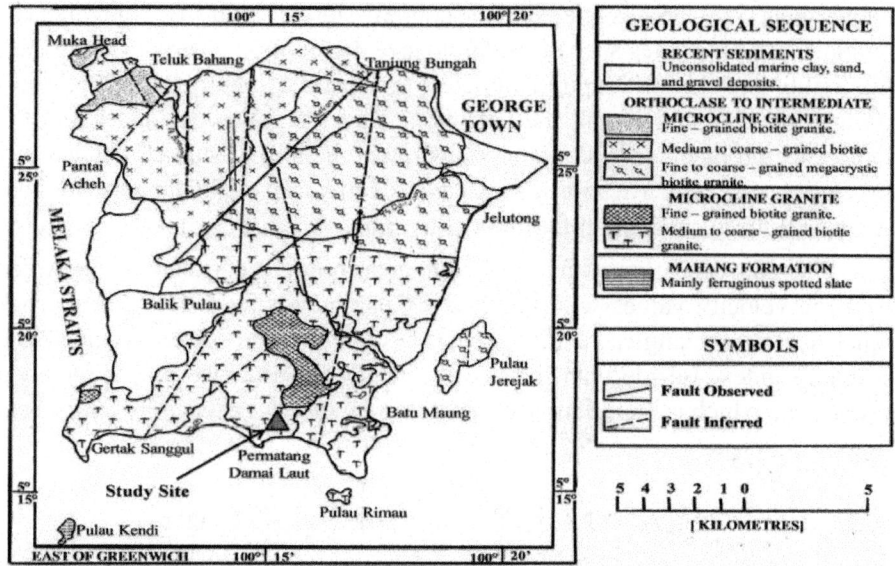

Figure 1. Geological map of first investigation site area (Penang, Malaysia).

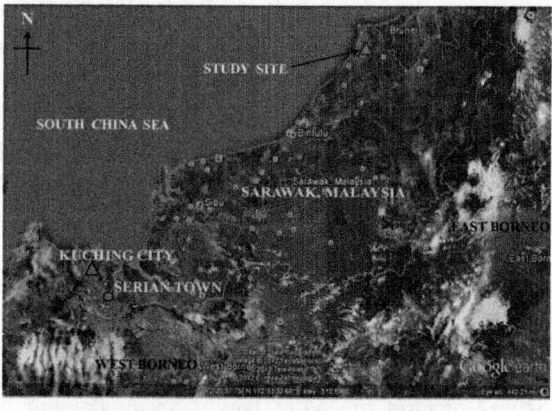

Figure 2. Second investigation site area (Sarawak, Malaysia).

MATERIALS AND METHODS APPLIED

In geosciences field, researchers have been used geophysical methods to study the Earth's subsurface structure and condition. Recently, determination and monitoring of aquifer formations have been done using geophysical survey of geoelectric soundings [10] and combination of electrical resistivity and engineering soil parameters to study the clayey sand soil's behaviour [11]. Beside than that, other researchers have used the combination of Self

Potential (SP), Electromagnetic (EM) and Resistivity profiling methods in the delineation of conductive zones [12].

However in engineering seismology, early all work is based upon the P-wave. P-waves are a type of elastic wave, also called seismic waves that can travel through gases (as sound waves), solids and liquids including the Earth. P-waves are produced by earthquakes and recorded by seismographs. The name P-wave is often said to stand either for primary wave, as it has the highest velocity and is therefore the first to be recorded; or pressure wave, as it is formed from alternating compressions and rarefactions. In isotropic and homogeneous solids, the mode of propagation of a P-wave is always longitudinal; thus, the particles in the solid have vibrations along or parallel to the travel direction of the wave energy. The velocity of P-waves in a homogeneous isotropic medium is given by Equation (1):

$$V_P \equiv \sqrt{\frac{K + \frac{4}{3}\mu}{\rho}} = \sqrt{\frac{\lambda + 2\mu}{\rho}} \qquad (1)$$

where K is the Hbulk modulusH (the modulus of incompressibility), μ is the Hshear modulusH (modulus of rigidity, sometimes denoted as G and also called the second H Lamé parameter H), ρ is the H density H of the material through which the wave propagates, and λ is the first H Lamé parameter H.

Of these, density shows the least variation, so the velocity is mostly controlled by K and μ. The H elastic modules H H P-wave modulus H, M, is defined so that $M = K + 4\mu/3$ and thereby it rewritten in Equation (2) as

$$V_P \equiv (M/\rho) \qquad (2)$$

Seismic imaging directs an intense sound source into the ground to evaluate subsurface conditions and to possibly detect high concentrations of contamination. Receivers called geophones, analogous to microphones, pick up "echoes" that come back up through the ground and record the intensity and time of the "echo" on computers. Data processing turns these signals into images of the geologic structure. This technology is similar in principle to active H electromagnetic survey H technology. For this seismic imaging, there are a few equipment is used to obtain the data from the field. The equipments involved in this seismic survey are a battery pack, a roll of trigger cable, two seismic cables, recording equipment (ABEM Terraloc Mark 8), 24 geophones of 14 Hertz, a striker metal plate and a log book. Terraloc Mark 8 is placed at the centre of seismic line and was connected to two seismic cables. 12 pounds hammer is used in this study in purpose of to safe the cost compare to gun. In this seismic survey, there 7 numbers of shot points with total of 168 traces recorded and 5.0 meter geophone interval was selected. This geophysical method will able to give a better analysis of the subsurface especially subsurface structure changes. Seismic refraction velocities in field are being compared with the higher frequency, typically ultrasonic measurements of the laboratory. The purpose is to determine their velocity index.

Rock-Quality Designation Method

We are also using RQD method from borehole samples. Rock-quality designation (RQD) is a rough measure of the degree of jointing or fracture in a rock mass, measured as a percentage of the drill core in lengths of 10 cm or more Rock quality designation (RQD) has several definitions. It is the borehole core recovery percentage incorporating only pieces of solid core that are longer than 100 mm in length measured along the centreline of the core. In this respect pieces of core that are not hard and sound should not be counted though they are 100 mm in length. RQD was originally introduced for use with core diameters of 54.7 mm (NX-size core). RQD has considerable value in estimating support of rock tunnels. RQD forms a basic element in some of the most used rock mass classification systems: Rock Mass Rating system (RMR) and Q-system.

RQD is defined in Equation (3) as the quotient [13]:

$$RQD = \left(\frac{l_{sum. of 100}}{l_{tot. core run}} \right) \times 100\%$$

(3)

where,

$l_{sum. of 100}$ = Sum of length of core sticks longer than 100 mm measured along the centre of the core.

$l_{tot. core run}$ = Total length of core run.

Standard Penetration Test (SPT) Method

Sampling from standard penetration test (SPT) can be either undisturbed, of which in-situ testing is a form, or disturbed. The principal sampling methods used in boreholes are:

SPT test is a dynamic test as described in BS1377 and is a measure of the density of the soil. The test incorporates a small diameter tube with a cutting shoe known as the "split barrel sampler" of about 650 mm length, 50 mm external diameter and 35 mm internal diameter. The sampler is forced into the soil dynamically using blows from a 63.5 kg hammer dropped through 760 mm. The sampler is forced 150 mm into the soil then the number of blows required to lower the sampler each 75 mm up to a depth of 300 mm is recorded. This is known as the "N" value. For coarse gravels the split barrel is replaced by a 60 degree cone.

There are some limitations for these tests. Firstly, it is affected by borehole disturbance, such as piping, base heave and stress relief. Second, it is affected by equipment to make borehole and by operator. Lastly, many corrections are required for interpretation and design. For that reason, in this study we carried out other applicable method to reduce the limitations which could be able to give the subsurface condition and information related to standard penetration test. For our first effort, in this study we try to use seismic refraction method to determine the correlation in between these two methods for environmental engineering applications.

RESULTS AND DISCUSSION

In this study, we used seismic refraction survey to locate the exact value of seismic velocities which can be correlated with the N values and RQD values from borehole data set. This seismic tomography result is processed using SeisOpt software. Our study not only focussed on correlation between seismic velocity analysis wit rock quality and soil's N values, however we also extended our study to soil's friction angle, soil's relative density and soil's compression strength. The main purposes we did this entire are to maximize the information of samples collected from the boreholes from both study areas. Perhaps, we able to identified the engineering characterizations of soil and rock samples from the boreholes and tie it with seismic P-wave velocity at the same depth locations.

For correlation between rock quality and seismic velocity analysis, we split the methods used into two parts. The first part is velocity determination from the seismic refraction survey and the second part is laboratory analysis for rock samples to determine their P-wave velocity. From the obtained results, then we found out the correlation for seismic velocity analysis and rock quality. Beside than that, we also study the correlation between seismic P-wave velocity changes due to their penetration strength. The seismic tomography used in this first study is 3 seismic lines which consisted of 4 borehole data sets. The seismic tomography for Line 01, Line 02 and Line 06 together with boreholes location is shown in Figures 3-5 in the next page.

Second site case is medium depth of investigation of the subsurface. The study area is located at Sarawak, Malaysia. The seismic refraction survey was conducted at the study site together with the geotechnical engineering practices in purpose to obtain the engineering parameters same as first site. There are 5 numbers of boreholes located along the seismic refraction survey line. The difference second case compared with the first case is medium depth of investigation (25 m to 40 m from the surface) and recorded N value greater than 50% of the average is counted into our consideration.

After analysing all the collected soil and rock sample from our borehole, we figure out correlation between their characterizations. Empirical correlation between N value internal friction, φ' and relative density D_r values of cohesionless soil are available from this study. **Table 1** below gives the correlation of N values with the properties of granular soils.

It is found that the mathematical correlations exist between friction angel, φ' and relative density, D_r is found as Equations (4) and (5) below.

$$φ› = 25° + 0.15D_r$$

with fines > 5%(4)

$$φ› = 30° + 0.15D_r \text{ with fines} > 5\%(5)$$

N values are also correlated with the unconfined compression strength of cohesive soils as shown in**Table 2**. However, these correlations are not used in the proportioning of foundations. They can at best use as approximate pointers to the consistency and shear strength of cohesive soils.

Table 1. Correlations with N values of cohesionless soils.

N Value	Friction Angle, φ' (Deg.)	Relative Density, D_r (%)	Description
Less than 4	25 - 28	Less than 15	Very loose
4 - 10	29 - 32	15 - 60	Loose
10 - 30	33 - 35	60 - 75	Medium
30 - 50	36 - 40	75 - 90	Dense
Over 50	41 - 45	Over 90	Very dense

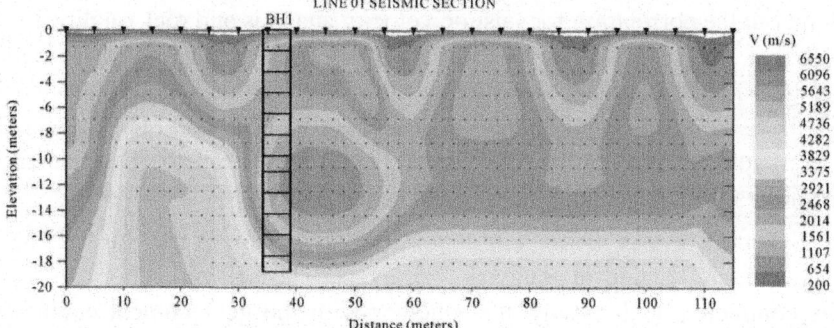

Figure 3. Line 01 seismic refraction section.

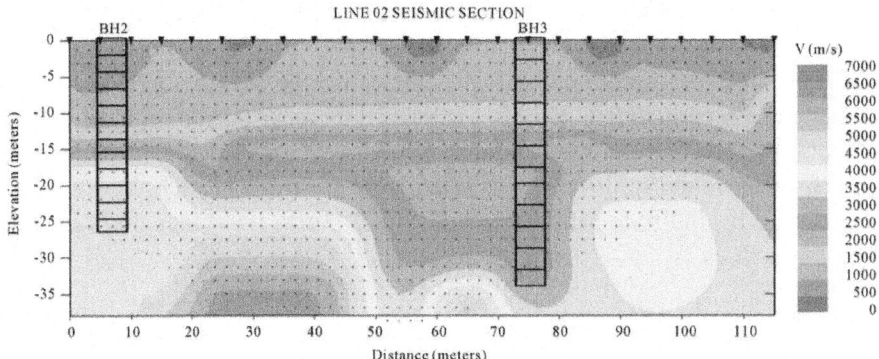

Figure 4. Line 02 seismic refraction section.

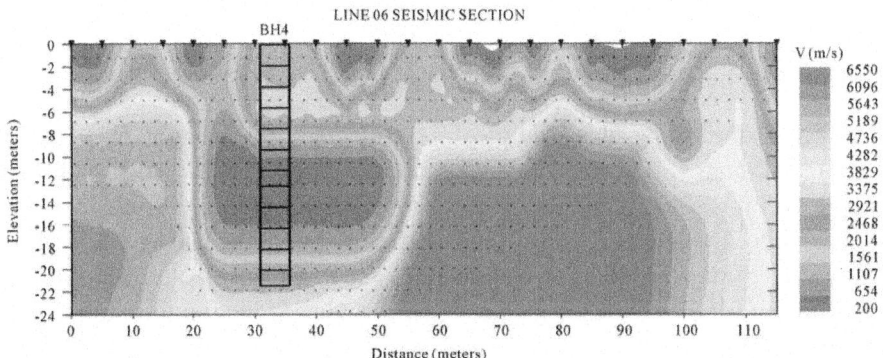

Figure 5. Line 06 seismic refraction section.

Table 2. Correlation with N values for cohesive soils.

N Value	Unconfined Compression Strength (kg·cm⁻²)	Consistency
Less than 2	Less than 0.25	Very soft
2 - 5	0.25 - 0.50	Soft
5 - 9	0.50 - 1.00	Medium
9 - 17	1.00 - 2.00	Stiff
17 - 33	2.00 - 4.00	Very stiff
Over 33	Over 4.00	Hard

RQD value is defined as the percentage of core that has core sticks greater than 10 cm long, for selected structural domains, or for specific length of core.

The **Table 3** and **Figure 6** show the central trend of this relationship, which however shows considerable scatter. It is found that the ratio (V_{FIELD}/V_{LAB}) when squared, was numerically close to the value of RQD (expressed as a ration rather than a percentage. This is applicable for near surface measurements. Equation (6) below was then comparing with previous study such as [14] and the equation found in this study is acceptable for tropical rock. However, the stress face by the rock samples would affect the RQD values and also their velocity values.

$$RQD(\%) = 0.97 \left[\frac{V_{FIELD}}{V_{LAB}} \right]^2 (100)$$

which is nearly to

$$RQD(\%) \approx \left[\frac{V_{FIELD}}{V_{LAB}} \right]^2 (100) \tag{6}$$

where V_F is field value of V_p and V_L is laboratory value of V_p.

From our study sites, samples are collected and brought back for the laboratory tests. From the laboratory analysis, the typical values of friction angle, φ' for some granular soils and silts are given in **Table 4** below. Meanwhile in **Figure 7** shows the location of boreholes at the study site in Sarawak, Malaysia. The correlation between P-wave velocity (infield) with N value and RQD for Sarawak study site also determined in this study site as shown in **Table 5**.

In this study, we also study the empirical correlation between Standard Penetration Test (N values), Rock Quality Designation values and infield seismic P-wave velocity values as shown in **Figure 8**. We found that the empirical correlation for tropical environmental study is $V_p = 23.605(N) - 160.43$ and the regression found is 0.9315 (93.15%). Meanwhile, the empirical correlation between P-wave velocities and RQD values is found as $V_p = 21.951(RQD) + 0.1368$ and the regression found is 0.8377 (83.77%).

From both selected study sites (Penang and Sarawak site), we were also determined correlation between apparent P-wave velocities and penetration strength. Penetration strength shows a strong influence on P-wave velocity for tropical granitic rock. **Figure 9** shows the empirical correlation between apparent P-wave velocities with penetration strength values for the study site. The corelation found as $V_{P(APP)} = 292.59(Ps) + 474.69$ and the regression coefficient is found as 0.9756.

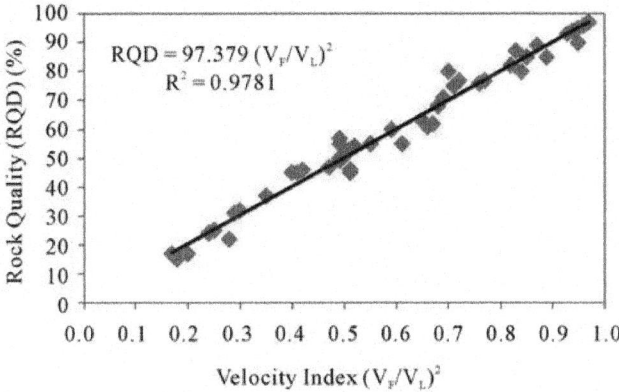

Figure 6. The trend of relationship between rock quality and velocity index.

Table 3. Relationship between RQD, velocity index and N value.

Quality Description	RQD (%)	Velocity Index $(V_F/V_L)^2$	N Value
Very poor	Less than 25	0 - 0.25	50 - 65
Poor	25 - 50	0.25 - 0.53	65 - 70
Fair	50 - 75	0.53 - 0.75	70 - 75
Good	75 - 85	0.75 - 0.85	75 - 85
Excellent	Over 85	Over 0.85	Over 85

Table 4. Typical values of drained friction angle for mixes soils and silts.

Soil type	Friction angle, φ' (deg)
Gravel mixed with sand	33 - 47
Clayey sand soils	
Loose	17 - 25
medium	25 - 40
Dense	40 - 55
Fine	
Silts	27 - 35

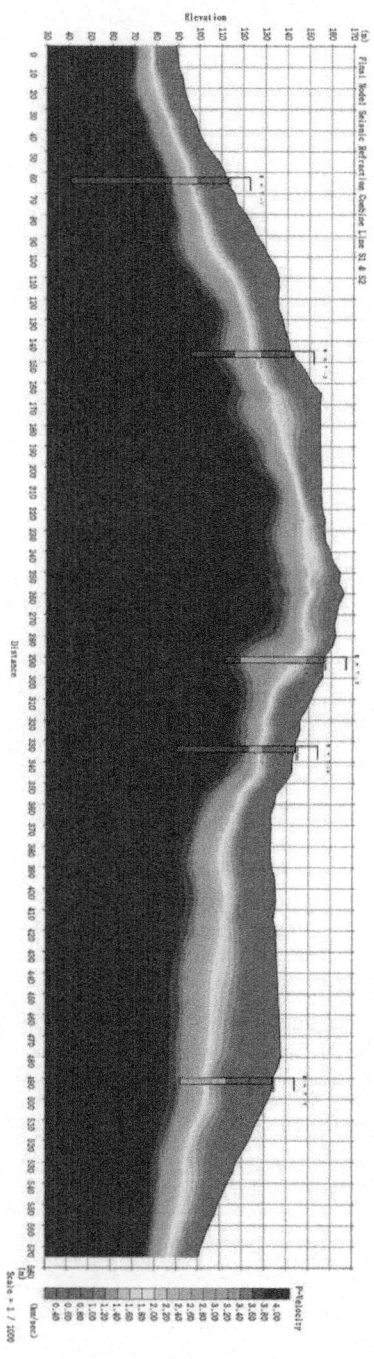

Figure 7. Seismic refraction tomography section at Sarawak, Malaysia (Second site).

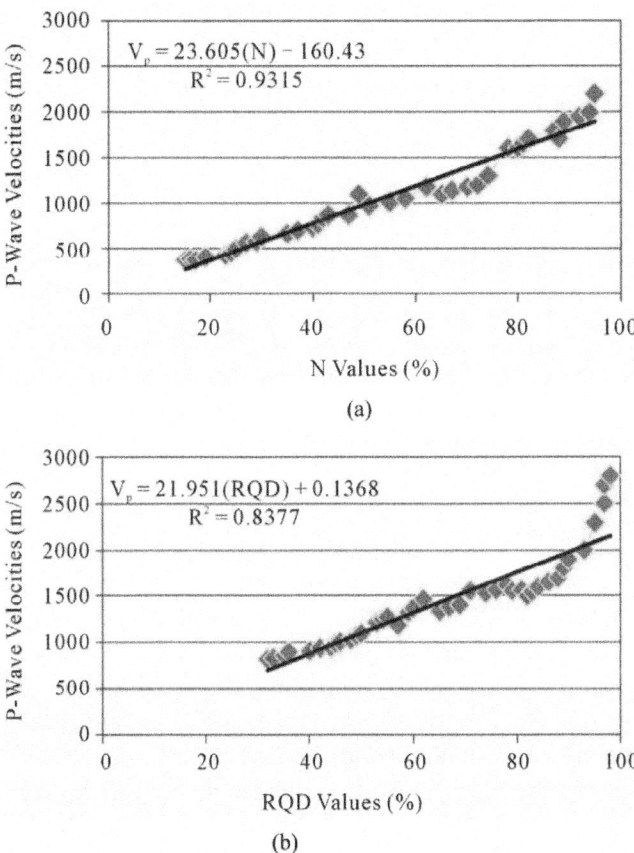

Figure 8. Empirical correlation of (a) P-wave velocities with N values and (b) P-wave velocities with RQD values for both studied areas.

Table 5. Correlation between P-wave velocity (infield) with N value and RQD for medium depth investigation.

Geological Classification	P-Wave Velocity (km/sec)	N Value (%)	RQD Value (%)
Residual Soils	0.4 - 1.0	Less than 50	-
Completely Weathered Granite	1.0 - 1.7	50 - 65	Less than 50
Highly Weathered Granite	1.7 - 2.1	65 - 75	50 - 70
Moderately Weathered Granite	2.1 - 2.7	75 - 85	70 - 85
Fresh Granite	Over 2.7	Over 85	Over 85

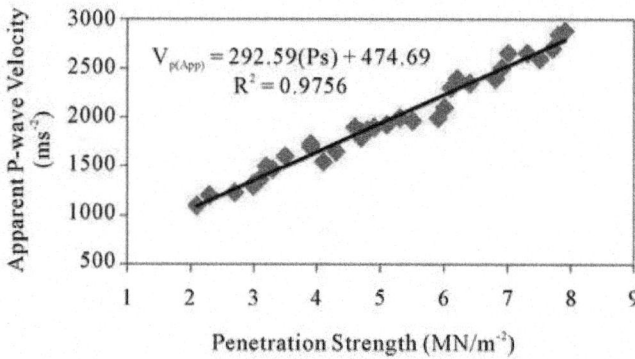

Figure 9. Correlation between apparent P-wave velocities with penetration strength values for tropical granitic rock (average from Penang and Sarawak site).

CONCLUSION

From the both selected study sites above, the conclusion can be made based on the results obtain. Firstly the objective of this study is successfully achieved. For many environmental site investigation studies, it has proved to be a very cost-effective method, especially when combined with other geophysical or geotechnical surveys. In our study, the estimation of engineering parameters from P-wave seismic velocity values is applicable for tropical environmental study. The estimation can help in predict the engineering characterizations and seismic P-wave velocity of the subsurface material (soils and rocks). In this study, the selected areas (Penang and Sarawak site) are both have same type of bedrock (Tropical Granitic Bedrock). This study however, can be trying for other areas with different type of weather condition and latitude region. The different weather and latitude also influent the weathering rate of rock. Thus the parameters obtain will influent by the surrounding condition.

ACKNOWLEDGEMENTS

I would like to wish thank Jeff Steven and Eva Diana for giving suggestion for this research. I also wish to thank anonymous reviewers for insightful comments that helped improved the quality of the manuscript.

REFERENCES

1.   A. A. Bery and R. Saad, "Clayey Sand Soil's Behaviour Analysis and Imaging Subsurface Structure via Engineering Characterizations and Integrated Geophysicals Tomography Modeling Methods," International Journal of Geosciences, Vol. 3, No. 1, 2012, pp. 93-104. Hdoi:10.4236/

ijg.2012.31011

2. S. A. Naeini and M. H. Baziar, "Effect of Fines Content on Steady-State Strength of Mixed and Layered Samples of Sand," Soil Dynamics and Earthquake Engineering, Vol. 24, No. 3, 2004, pp. 181-187. Hdoi:10.1016/j.soildyn.2003.11.003

3. B. O. Hardin, "1-D Strain in Normally Consolidated Cohesionless Soils," Journal of Geotechnical Engineering Division, Vol. 113, No. 12, 1987, pp. 1449-1467. Hdoi:10.1061/(ASCE)0733-9410(1987)113:12(1449)

4. F. A. Chuhan, A. Kjeldstad, K. Bjorlykke and K. Hoeg, "Experimental Compression of Loose Sands: Relevance to Porosity Reduction during Burial in Sedimentary Basins," Canadian Geotechnical Journal, Vol. 40, 2003, pp. 995-1011. Hdoi:10.1139/t03-050

5. F. David, "Essentials of Soil Mechanics and Foundations Basic Geotechnics," 6th Edition, Pearson Education, Upper Saddle River, 2007.

6. B. Vásárhelyi and P. Ván, "Influence of Water Content on the Strength of Rock," Engineering Geology, Vol. 84, No. 1, 2006, pp. 70-74. Hdoi:10.1016/j.enggeo.2005.11.011

7. M. Romana and B. A. Vásárhelyi, "A Discussion on the Decrease of Unconfined Compressive Strength between Saturated and Dry Rock Samples," Polytechnic University of Valencia, Valencia, 2007.

8. W. S. Ong, "The Geology and Engineering Geology of Penang Island," Geological Survey of Malaysia, 1993.

9. H. S. Abieda, Z. Z. T. Harith and A. H. A. Rahman, "Depositional Controls on Petrophysical Properties and Reservoir Characteristics of Middle Miocene Miri Formation Sandstones, Sarawak," Bulletine of the Geological Society of Malaysia, Vol. 5, 2005, pp. 63-75.

10. L. Ouadif, L. Bahi, A. Akhssas, K. Baba and M. Menzhi, "Geophysics Contribution for the Determination of Aquifers with a Case Study," International Journal of Geosciences, Vol. 3, No. 1, 2012, pp. 117-125. doi:10.4236/ijg.2012.31014

11. A. Bery and R. Saad, "Tropical Clayey Sand Soil's Behaviour Analysis and Its Empirical Correlations via Geophysics Electrical Resistivity Method and Engineering Soil Characterizations," International Journal of Geosciences, Vol. 3, No. 1, 2012, pp. 111-116. doi:10.4236/ijg.2012.31013

12. O. T. Nkereuwem, S. N. Yusuf and M. U. Mijinyawa, "An Integration of Self Potential, Electromagnetic and Resistivity Profiling Methods in the Search for Sulfide Deposits in Gwoza, Borno State, Nigeria," International Journal of Geosciences, Vol. 3, No. 2, 2012, pp. 365-372.

doi:10.4236/ijg.2012.32040

13. D. U. Deere and D. W. Deere, "The Rock Quality Designation (RQD) Index in Practice," In: L. Kirkaldie, Ed., Rock Classification System for Engineering Purposes, ASTM STP 984, American Society for Testing and Materials, Philadelphia, 1988, pp. 91-101.doi:10.1520/STP48465S

14. D. U. Deere, A. J. Hendron, F. D. Patton and E. J. Cording, "Design of Surface and Near Surface Construction in Rock. In Failure and Breakage of Rock," Proceedings of 8th US Symposium Rock Mechanics, Society of Mining Engineers, American Institute of Mining, Metallurgical and Petroleum Engineers (SAUS), New York, 1967, pp. 237-302.

Chapter 5

STRESS/STRAIN-DEPENDENT PROPERTIES OF HYDRAULIC CONDUCTIVITY FOR FRACTURED ROCKS

Yifeng Chen and Chuangbing Zhou

State Key Laboratory of Water Resources and Hydropower Engineering Science, Key Laboratory of Rock Mechanics in Hydraulic Structural Engineering, Wuhan University, P. R. China

INTRODUCTION

In the last two decades there has seen an increasing interest in the coupling analysis between fluid flow and stress/deformation in fractured rocks, mainly due to the modeling requirements for design and performance assessment of underground radioactive waste repositories, natural gas/oil recovery, seepage flow through dam foundations, reservoir induced earthquakes, etc. Characterization of hydraulic conductivity for fractured rock masses, however, is one of the most challenging problems that are faced by geotechnical engineers. This difficulty largely comes from the fact that rock is a heterogeneous geological material that contains various natural fractures of different scales (Jing, 2003). When engineering works are constructed on or in a rock mass, deformation of both the fractures and intact rock will usually occur as a result of the stress changes. Due to the stiffer rock matrix, most deformation occurs in the fractures, in the form of normal and shear displacement. As a result, the existing fractures may close, open, grow and new fractures may be induced, which in turn changes the structure of the rock mass concerned and alters its fluid flow behaviors and properties. Therefore, the fractures often play a dominant role in understanding the flow-stress/deformation coupling behavior of a rock system, and their mechanical and hydraulic properties have to be properly established (Jing, 2003).

Traditionally, fluid flow through rock fractures has been described by the cubic law, which follows the assumption that the fractures consist of two smooth parallel plates. Real rock fractures, however, have rough walls, variable aperture and asperity areas where the two opposing surfaces of the fracture walls are in contact with each other (Olsson & Barton, 2001). To simplify the problem, a single, average value (or together with its stochastic characteristics) is commonly used to describe the mechanical aperture of an individual fracture. A great amount of work (Lomize, 1951; Louis, 1971; Patir & Cheng, 1978; Barton et al., 1985; Zhou & Xiong, 1996) has been done to find an equivalent, smooth wall hydraulic aperture out of the real mechanical aperture such that when Darcy's law or its modified version is applied, the equivalent smooth fracture yields the same water conducting capacity with its original rough fracture. It is worth noting that clear distinction manifests between the geometrically measured mechanical aperture (denoted by b in the context) and the theoretical smooth wall hydraulic aperture (denoted by b^*), and the former is usually larger in magnitude than the latter due to the roughness of and filling materials in rock fractures (Olsson & Barton, 2001).

The ubiquity of fractures significantly complicates the flow behavior in a discontinuous rock mass. The primary problem here is how to model the flow system and how to determine its corresponding hydraulic properties for flow analysis. Theoretically, the representative elementary volume (REV) of a rock mass can serve as a criterion for selecting a reasonable hydromechanical model. This statement relates to the fact that REV is a fundamental concept that bridges the micro-macro, discrete-continuous and stochastic-determinate behaviors of the fractured rock mass and reflects the size effect of its hydraulic and mechanical properties. The REV size for the hydraulic or mechanical behavior is a macroscopic measurement for which the fractured medium can be seen as a continuum. It is defined as the size beyond which the rock mass includes a large enough population of fractures and the properties (such as hydraulic conductivity tensor and elastic compliance tensor) basically remain the same (Bear, 1972; Min & Jing, 2003; Zhou & Yu, 1999; Wang & Kulatilake, 2002). Owing to high heterogeneity of fractured rock masses, however, the REV can be very large or in some situations may not exist. If the REV does not exist, or is larger than the scale of the flow region of interest, it is no longer appropriate to use the equivalent continuum approach. Instead, the discrete fracture flow approach may be applied to investigate and capture the hydraulic behavior of the fractured rock masses. However, due to the limited available information on fracture geometry and their connectivity, it is not a trivial task to make a detailed flow path model. Thus, in practice, the equivalent continuum model is still the primary choice to approximate the hydraulic behavior of discontinuous rocks.

The hydraulic conductivity tensor is a fundamental quantity to characterizing the hydromechanical behavior of a fractured rock. Various techniques have been proposed to quantify the hydraulic conductivity tensor, based on results from field tests, numerical simulations, and back analysis techniques, etc. Earlier investigations focused on using field measurements (e.g. aquifer pumping test or packer test (Hsieh & Neuman, 1985)) to estimate the three-dimensional hydraulic conductivity tensor. This approach, however, is generally time-consuming, expensive and needs well controlled experimental conditions. Numerical and analytical methods are also used to estimate the hydraulic properties of complex rock masses due to its flexibility in handling variations of fracture system geometry and ranges of material properties for sensitivity or uncertainty estimations. In the literature, both the equivalent continuum approach (Snow, 1969; Long et al., 1982; Oda, 1985; Oda, 1986; Liu et al., 1999; Chen et al., 2007; Zhou et al., 2008) and the discrete approach (Wang & Kulatilake, 2002; Min et al., 2004) are widely applied. In this chapter, however, only the equivalent continuum approach is focused for its capability of representing the overall behavior of fractured rock masses at large scales.

Among many others, Snow (1969) developed a mathematical expression for the permeability tensor of a single fracture of arbitrary orientation and aperture and considered that the permeability tensor for a network of such fractures can be formed by adding the respective components of the permeability tensors for each individual fracture. Oda (1985, 1986) formulated the permeability tensor of rock masses based on the geometrical statistics of related fractures. Liu et al. (1999) proposed an analytical solution that links changes in effective porosity and hydraulic conductivity to the redistribution of stresses and strains in disturbed rock masses. Zhou et al. (2008) suggested an analytical model to determine the permeability tensor for fractured rock masses based on the superposition principle of liquid dissipation energy. Although slight discrepancy exists between the permeability tensor and the hydraulic conductivity tensor (the former is an intrinsic property determined by fracture geometry of the rock mass, while the latter also considers the effects of fluid viscosity and gravity), when taking into account the flow-stress coupling effect, the above models presented, respectively, by Snow (1969), Oda (1985) and Zhou et al. (2008) were proved to be functionally equivalent for a certain fluid (Zhou et al., 2008). A common limitation with the above models lies in the fact that the hydraulic conductivity tensor of a fractured rock mass is all formulated to be either stress-dependent or elastic strain-dependent. Consequently, material nonlinearity and post-peak dilatancy are not considered in the formulation of the hydraulic conductivity tensor for disturbed rock masses. To address this problem, Chen et al. (2007) extended the above work and proposed a numerical model to establish the hydraulic conductivity for fractured rock masses under complex loading conditions. Based on the observation that natural fractures in a rock mass are most often clustered in certain critical orientations resulting from their geological modes and history of formation (Jing, 2003), characterizing the rock mass as an equivalent continuum containing one or multiple sets of planar and parallel fractures with various critical orientations, scales and densities turns out to be a desirable approximation. Starting from this point of view, the deformation patterns of the fracture network can be first characterized by establishing an equivalent elastic or elasto-plastic constitutive model for the homogenized medium. On this basis, a stress-dependent hydraulic conductivity tensor may be formulated for the former for describing the hydraulic behavior of the rock mass at low stress level and with overall elastic response; and a strain-dependent hydraulic conductivity tensor for the latter for demonstrating the influences of material non-linearity and shear dilatancy on the hydraulic properties after post-peak loading. This chapter mainly presents the research results on the stress/strain-dependent hydraulic properties of fractured rock masses under mechanical loading or engineering disturbance achieved by Chen et al. (2006), Zhou et al. (2006), Chen et al. (2007) and Zhou et al. (2008). The stress-dependent hydraulic conductivity model (Zhou et al., 2008) was proposed for estimation of the hydraulic properties of fractured rock masses at relatively lower stress level based on the superposition principle of flow dissipation energy. It was shown that the model is equivalent to Snow's model (Snow, 1969) and Oda's model (Oda, 1986) not only in form but also in function when considering the effects of mechanical loading process on the evolution of hydraulic properties. This model relies on the geometrical characteristics of rock fractures and the corresponding fracture network, and demonstrates the coupling effect between fluid flow and deformation. In this model, the pre-peak dilation and contraction effect of the fractures under shear loading is also empirically considered. It was applied to estimate the hydraulic properties of the rock mass in the dam site of the Laxiwa Hydropower Project located in the upstream of the Yellow River, China, and the model predictions have a good agreement with the site observations from a large number of singlehole packer tests.

The strain-dependent hydraulic conductivity model (Chen et al., 2007), on the other hand, was established by an equivalent non-associative elastic-perfectly plastic constitutive model with mobilized dilatancy to characterize the nonlinear mechanical behavior of fractured rock masses under complex loading conditions and to separate the deformation of weaker fractures from the overall deformation response of the homogenized rock masses. The major advantages of the model lie in the facts that the proposed hydraulic conductivity tensor is related to strains rather than stresses, hence enabling hydro-mechanical coupling analysis to include the effect of material nonlinearity and post-peak dilatancy, and the proposed model is easy to be included in a FEM code, particularly suitable for numerical analysis of hydromechanical problems in rock engineering with large scales. Numerical simulations were performed to investigate the changes in hydraulic conductivities of a cube of fractured rock mass under triaxial compression and shear loading as well as an underground circular excavation in biaxial stress field at the Stripa mine (Kelsall et al., 1984; Pusch, 1989), and the simulation results are justified by in-situ experimental observations and compared with Liu's elastic strain-dependent analytical solution (Liu et al., 1999).

Unless otherwise noted, continuum mechanics convention is adopted in this chapter, i.e., tensile stresses are positive while compressive stresses are negative. The symbol (:) denotes an inner product of two second-order tensors (e.g., $a{:}b{=}a_{ij}b_{ij}$) or a double contraction of adjacent indices of tensors of rank two and higher (e.g., $c{:}d{=}c_{ijkl}d_{kl}$), and ($\otimes$) denotes a dyadic product of two vectors (e.g., $a{\otimes}b{=}a_i b_j$) or two second-order tensors (e.g., $c{\otimes}d{=}c_{ij}d_{kl}$).

STRESS-DEPENDENT HYDRAULIC CONDUCTIVITY OF ROCK FRACTURES

In this section, the elastic deformation behavior of rock fractures at the pre-peak loading region will be first presented, and then a stress-dependent hydraulic conductivity model will be formulated. The deformation model (or indirectly the hydraulic conductivity model) is validated by the laboratory shear-flow coupling test data obtained by Liu et al. (2002). The main purpose of this section is to provide a theory for developing a stress-dependent hydraulic conductivity tensor for fractured rock masses that will be presented later in Section 4.

Characterization of Rock Fractures

One of the major factors that govern the flow behavior through fractured rocks is the void geometry, which can be described by several geometrical parameters, such as aperture, orientation, location, size, frequency distribution, spatial correlation, connectivity, and contact area, etc. (Olsson & Barton, 2001; Zhou et al., 1997; Zhou & Xiong, 1997). Real fractures are neither so solid as intact rocks nor void only. They have complex surfaces and variable apertures, but to make the flow analysis tractable, the geometrical description is usually simplified. It is common to assume that individual fractures lie in a single plane and have a constant hydraulic aperture. When the fractures are subjected to normal and shear loadings, the fracture aperture, the contact area and the matching between the two opposing surfaces will be altered. As a result, the equivalent hydraulic aperture of the fractures varies with their normal and shear stresses/displacements, which demonstrates the apparent coupling mechanism between fluid flow and stress/deformation (Min et al., 2004).

The aperture of rock fractures tends to be closed under applied normal compressive stress. The asperities of the surfaces will be crushed when their localized compressive stresses exceed their compressive strength. As a large number of asperities are crushed under high compressive stress, the contact area between the fracture walls increases remarkably and the crushed rock particles partially or fully fill the nearby void, which decreases the effective flow area, reduces the hydraulic conductivity of the fracture, and even changes the flow paths through fracture plane. Fig. 1 depicts the increase in contact area of fractures under increasing compressive stresses modelled by boundary element method (Zimmerman et al., 1991).

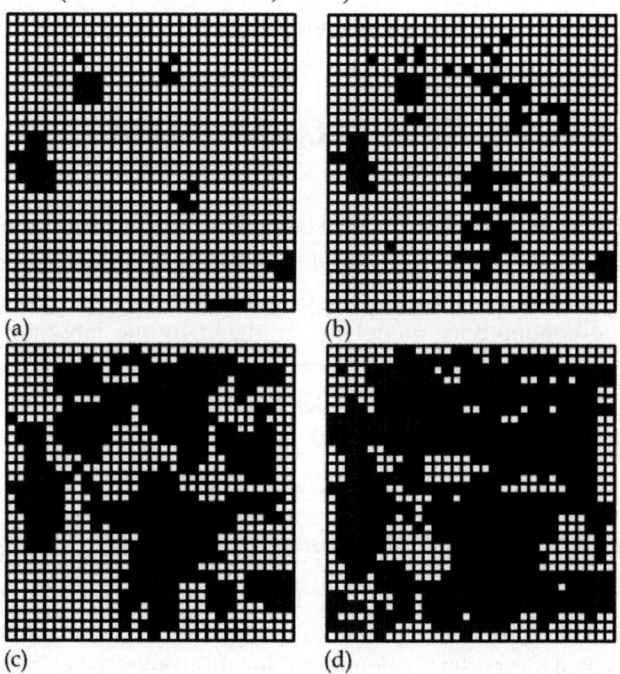

Figure 1. Variation of contact surface of fractures under increasing compressive stresses (after Zimmerman et al., (1991): (a) P=0 MPa; (b) P=20 MPa; (c) P=40 MPa and (d) P=60 MPa.

The coupling process between fluid flow and shear deformation is more related to the roughness of fractures and the matching of the constituent walls. Fig. 2 shows the impact of the fracture structure on the shear stress-deformation coupling mechanism. In Fig. 2(a), the opposing walls of the fracture are well matched so that the fracture always dilates and the hydraulic conductivity increases under shear loading as long as the applied normal stress is not high enough for the asperities to be crushed. For the state shown in Fig. 2(c), shear loading will result in the closure of the fracture and the reduction in hydraulic conductivity. Fig. 2(b) illustrates a middle state between (a) and (c), and its shearing effect depends on the direction of shear stress. When the matching of a fracture changes from (a) to (b) then to (c) under shear loading, shear dilation occurs. On the other hand, shear contraction takes place from the movement of the matching from (c) to (b) then to (a). In a more complex scenario, shear dilation and shear contraction may happen alternately, resulting in the fluctuation of the hydraulic behavior of the fractures.

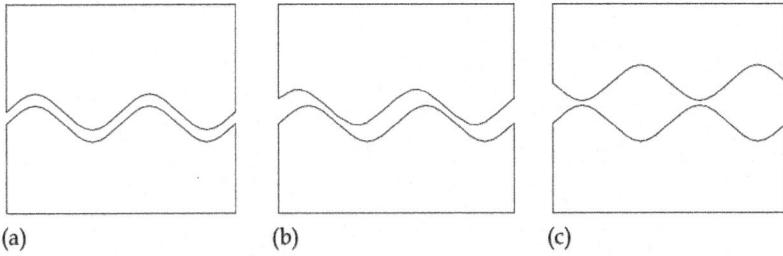

(a)   (b)   (c)

Figure 2. Shear dilation and shear contraction of fractures: (a) well-matched; (b) fair-matched; and (c) bad-matched.

An elastic constitutive model for rock fractures To formulate the stress-dependent hydraulic conductivity for rock fractures, we model the fractures by an interfacial layer, as shown in Fig. 3. The interfacial layer is a thin layer with complex constituents and textures (depending on the fillings, asperities and the contact area between its two opposing walls). Assumption is made here that the apparent mechanical response of the interfacial layer can be described by Lame's constant λ and shear modulus μ. Because the thickness of the interfacial layer (i.e., the initial mechanical aperture of the fracture) is generally rather small comparing to the size of rock matrix, it is reasonable to assume that $\varepsilon_x = \varepsilon_y = 0$ and $\gamma_{xy} = \gamma_{yx} = 0$ within the interfacial layer. Then according to the Hooke's law of elasticity, the elastic constitutive relation for the interfacial layer under normal stress σ_n and shear stress τ can be written in the following incremental form:

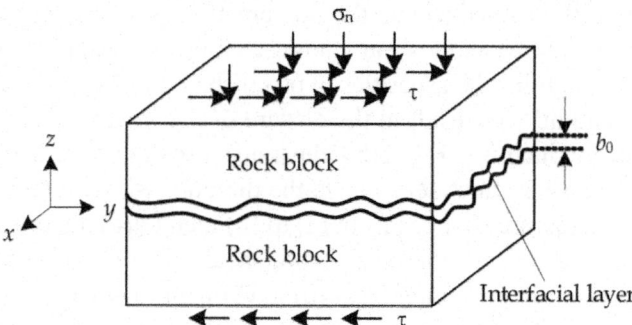

Figure 3. The interfacial layer model for rock fractures.

$$\begin{Bmatrix} d\sigma'_n \\ d\tau \end{Bmatrix} = \begin{bmatrix} \lambda + 2\mu & 0 \\ 0 & \mu \end{bmatrix} \begin{Bmatrix} d\varepsilon_n \\ d\gamma \end{Bmatrix}$$

$$(1)$$

For convenience, we use u_1 to denote the relative normal displacement of the interfacial layer caused by the effective normal stress σ'_n, δ to denote the relative tangential displacement caused by the shear stress τ, and u_2 to denote the relative normal displacement caused by shear dilation or contraction (positive for dilatant shear, negative for contractive shear). Hence, the total normal relative displacement u is represented as

$$u = u_1 + u_2$$

$$(2)$$

The increments of strains, $d\varepsilon_n$ and $d\gamma$, can be expressed in terms of the increments of relative displacements, du_1 and $d\delta$, as follows:

$$\begin{cases} d\varepsilon_n = du_1 / (b_0 + u) \\ d\gamma = d\delta / (b_0 + u) \end{cases}$$

$$(3)$$

where b_0 is the thickness of the interfacial layer or the initial mechanical aperture of the fracture. Substituting Eq. (3) in Eq. (1) yields:

$$\begin{Bmatrix} d\sigma'_n \\ d\tau \end{Bmatrix} = \begin{bmatrix} k_n & 0 \\ 0 & k_s \end{bmatrix} \begin{Bmatrix} du_1 \\ d\delta \end{Bmatrix}$$

$$(4)$$

where k_n and k_s denote the tangential normal stiffness and tangential shear stiffness of the interfacial layer, respectively.

$$k_n = (\lambda + 2\mu) / (b_0 + u), \quad k_s = \mu / (b_0 + u)$$

$$(5)$$

Interestingly, k_n and k_s show a hyperbolic relation with normal deformation and characterize the deformation response of the interfacial layer under the idealized conditions that each fracture is replaced by two smooth parallel planar plates connected by two springs with stiffness values k_n and k_s. As can be seen from Eq. (5), as long as the initial normal stiffness and shear stiffness with zero normal displacement, k_{n0} and k_{s0}, are known, they can be used as substitutes for λ and μ.

Substituting Eq. (2) in Eq. (4) results in:

$$d\sigma'_n = \frac{(\lambda + 2\mu)du_1}{b_0 + u_1 + u_2}$$

(6)

$$d\tau = \frac{\mu d\delta}{b_0 + u_1 + u_2}$$

(7)

Suppose normal stress σ_n is firstly applied before the loading of shear stress, u_1 can be obtained by directly integrating Eq. (6):

$$u_1 = (b_0 + u_2)\left[\exp\left(\frac{\sigma'_n}{\lambda + 2\mu}\right) - 1\right]$$

(8)

Here, it is to be noted that the elastic constitutive model for the rock fracture leads to an exponential relationship between the fracture closure and the applied normal stress, which has been widely revealed in the literature, e.g., in Min et al. (2004).

On the other hand, the shear expansion caused by $d\delta$ can be estimated from shear dilation angle d_m:

$$du_2 = \tan d_m \, d\delta$$

(9)

By introducing two parameters, s and φ, pertinent to normal stress σ_n, we represent the dilation angle d_m under normal stress σ_n in the form of Barton's strength criterion for joints (Barton, 1976) ($\tau = \sigma_n \tan(2d_m + \varphi_b)$, where φ_b is the basic frictional angle of joints):

$$\tan d_m = \frac{1}{2}\left[\arctan\left(\frac{\tau}{s}\right) - \varphi\right]$$

(10)

Obviously, s is a normal stress-like parameter, and φ is a frictional angle-like parameter. But to make the above formulation still valid into pre-dilation state (i.e., shear contraction state), s and φ differ from their initial implications. Later, we will show how they can be back calculated from shear experimental data. Substituting Eqs. (9) and (10) into (7) yields:

$$\frac{du_2}{b_0 + u_1 + u_2} = \frac{1}{2\mu}\left[arctan\left(\frac{\tau}{s}\right) - \varphi \right] d\tau$$

(11)

By integrating Eq. (11), we have:

$$u_2 = (b_0 + u_1)\left\{ \exp\left[\frac{|\tau|}{2\mu}\left(arctan\frac{|\tau|}{s} - \varphi \right) - \frac{s}{4\mu} \ln\left(1 + \frac{\tau^2}{s^2}\right) \right] - 1 \right\}$$

(12)

By solving the simultaneous equations, Eqs. (8) and (12), we have:

$$\begin{cases} u_1 = \dfrac{A(1+B)}{1-AB}b_0 \\ u_2 = \dfrac{B(1+A)}{1-AB}b_0 \end{cases}$$

(13)

where

$$A = exp\left(\frac{\sigma'_n}{\lambda + 2\mu} \right) - 1$$

(14)

$$B = exp\left[\frac{|\tau|}{2\mu}\left(arctan\frac{|\tau|}{s} - \varphi \right) - \frac{s}{4\mu}\ln\left(1 + \frac{\tau^2}{s^2}\right) \right] - 1$$

(15)

Thus, the total normal deformation under normal and shear loading can be obtained,

$$u = u_1 + u_2 = \frac{A + B + 2AB}{1 - AB}b_0$$

(16)

The actual aperture of the fracture, $b = b_0 + u$, is given by:

$$b = b_0 + u = (1 + \chi)b_0$$

(17)

where

$$\chi = \frac{A + B + 2AB}{1 - AB}$$

(18)

Stress-Dependent Hydraulic Conductivity for Rock Fractures

Since natural fractures have rough walls and asperity areas, it is not appropriate to directly use the aperture derived by Eq. (17) for describing the hydraulic conductivity of the fractures. Instead, an equivalent hydraulic aperture is usually taken to represent the percolation property of the fractures, as demonstrated in Section 1. Based on experimental data, the relationship between the equivalent hydraulic aperture and the mechanical aperture has been widely examined in the literature, and the empirical relations proposed by Lomize (1951), Louis (1971), Patir & Cheng (1978), Barton el al. (1985) and Olsson & Barton (2001) are listed in Table 1. For example, if Patir and Cheng's model is used to estimate the equivalent hydraulic aperture that accounts for the flow-deformation coupling effect in pre-peak shearing stage, then there is

$$b^* = (1 + \chi)b_0 \left[1 - 0.9\exp(-0.56 / C_v)\right]^{1/3} \tag{19}$$

where C_v is the variation coefficient of the mechanical aperture of the discontinuities, which is mathematically defined as the ratio of the root mean squared deviation to the arithmetic mean of the aperture. For convenience, Eq. (19) is rewritten as:

$$b^* = b_0 f(\beta) \tag{20}$$

Obviously, $f(\beta)$ is a function of the normal and shear loadings, the mechanical characteristics and the aperture statistics of the fractures.

Thus, the hydraulic conductivity of the fractures subjected to normal and shear loadings is approximated by the hydraulic conductivity of the laminar flow through a pair of smooth parallel plates with infinite dimensions:

$$k = \frac{g b^{*2}}{12v} \tag{21}$$

where k is the hydraulic conductivity, g is the gravitational acceleration, and v is the kinematic viscosity of the fluid.

An alternative approach to account for the deviation of the real fractures from the ideal conditions assumed in the parallel smooth plate theory is to adopt a dimensionless constant, ς, to replace the constant multiplier, $1/12$, in Eq. (21), where $0 < \varsigma < 1/12$ (Oda, 1986). In this manner, the hydraulic conductivity of the fractures is estimated by

$$k = \varsigma \frac{g b^2}{v} \tag{22}$$

Clearly, the constant, ς, approaches 1/12 with increasing scale and decreasing roughness of the fractures.

Eqs. (21) and (22) show that the hydraulic conductivity of a rock fracture varies quadratically with its mechanical aperture. The latter depends, by Eq. (18), on the normal and shear stresses applied on the fracture. Hence, we call the established model, Eq. (21) or (22), the stress-dependent hydraulic conductivity model, and it is suitable to describe the hydraulic behavior of the fractures subjected to mechanical loading in the pre-peak stage.

Table 1. Empirical relations between equivalent hydraulic aperture and mechanical aperture

Authors	Expressions	Descriptions
Lomize (1951)	$b^* = b\left[1.0 + 6.0(e/b)^{1.5}\right]^{-1/3}$	b^* is the equivalent hydraulic aperture of fractures, b the mechanical aperture, e the absolute asperity height, e_m the average asperity height, D_H the hydraulic radius, C_v the variation coefficient of the mechanical aperture, JRC the joint roughness coefficient, JRC_0 the initial value of JRC, JRC_{mob} the mobilized JRC, δ the shear displacement and δ_p the peak shear displacement.
Louis (1971)	$b^* = b\left[1.0 + 8.8(e_m/D_H)^{1.5}\right]^{-1/3}$	
Patir & Cheng (1978)	$b^* = b\left[1 - 0.9\exp(-0.56/C_v)\right]^{1/3}$	
Barton, et al. (1985)	$b^* = b^2 JRC^{-2.5}$	
Olsson & Barton (2001)	$\begin{cases} b^* = b^2 JRC_0^{-2.5} & \delta \le 0.75\delta_p \\ b^* = b^{1/2} JRC_{mob} & \delta \ge \delta_p \end{cases}$	

Validation of the Elastic Constitutive Model

The key point of the stress-dependent hydraulic conductivity model is whether the established elastic constitutive model can properly describe the variation of mechanical aperture under normal and shear loadings at low stress level. Here, we use the results of the laboratory test performed by Liu et al. (2002) to validate the mechanical model. The test was conducted to study shear-flow coupling properties for a marble fracture with fillings of sand under low normal stresses and small shear displacements.

The marble specimen for shear-flow coupling test is illustrated in Fig. 4, which was collected from the Daye Iron Mine in China. The uniaxial compressive strength and density of the rock sample are 52.4 MPa and 2.66×10^3 kg/m³, respectively. The specimen was cut into round shape and the fracture surfaces were polished, with its size of 290 mm in diameter and 200 mm in height. The opposite walls of the fracture were cemented with a layer of filtered sands with their diameters ranged from 0.5 to 0.69 mm, and the fracture was further filled with the same sands. The initial aperture of the fracture, b_0, is about 1.31 mm.

The coupled shear-flow test were conducted by first applying a prescribed normal stress ranging between 0.1 and 0.5 MPa and then applying shear displacement in steps until a maximum displacement of about 0.4 mm was reached. During tests, steady-state fluid flow rate and normal displacement were continuously recorded.

With such low normal stresses and small shear displacements, it is reasonable to consider that the fracture behaves elastic during the coupled shear-flow test. According to the experimental results, the elastic parameters, λ and μ, of the fracture with fillings are estimated as λ=1.81 MPa and μ=3.62 MPa. In order to enable Eq. (16) to predict the mechanical aperture of the facture under normal and shear loads, the normal stress-like parameter, s, and the frictional angle-like parameter, φ, should be further determined. Fortunately, both of them can be derived by fitting the experimental curve between normal displacement and shear displacement, as plotted in Fig. 5, using Eq. (16) such that the least square error is minimized. With this approach, we obtain that for σ_n=0.1 MPa, s=0.062, φ=1.324, and for σ_n=0.4 MPa, s=0.046, φ=1.310.

Fig. 5 plots the experimental results as well as the model predictions of the relation between mechanical aperture and shear displacement of the fracture under constant normal stresses. Generally, the proposed elastic constitutive model manifests the behavior of the fracture with fillings during the shear-flow coupling test with low normal and shear loads. Shear contraction is observed in the initial 0.06-0.08 mm of shear displacement, which is followed by shear dilation in the remaining of the shear displacement. This property, which is actually ensured by the empirical relation assumed in Eq. (9), suggests that the resultant model is suitable for phenomenologically describing the pre-peak shear-flow coupling effect of fractures.

Fig. 6 further depicts the sensitivity of s and φ on the behavior of the fracture. In Fig. 6(a), φ is fixed to 1.324, while s varies from 0.02 to 0.08. As s increases, shear contraction more apparently manifests, and the mechanical aperture versus shear displacement curves become lower as a whole. On the other hand, the effect of varying φ from 0.524 to 1.222 but fixing s to 0.062 is plotted in Fig. 6(b). For small value of φ, shear contraction is trivial and the curve extends with a larger slope. As φ increases, however, shear contraction becomes relatively remarkable and the curve turns relatively flat. Thus, by adjusting s and φ, the mechanical and hydraulic behaviors of the fracture can be appropriately established.

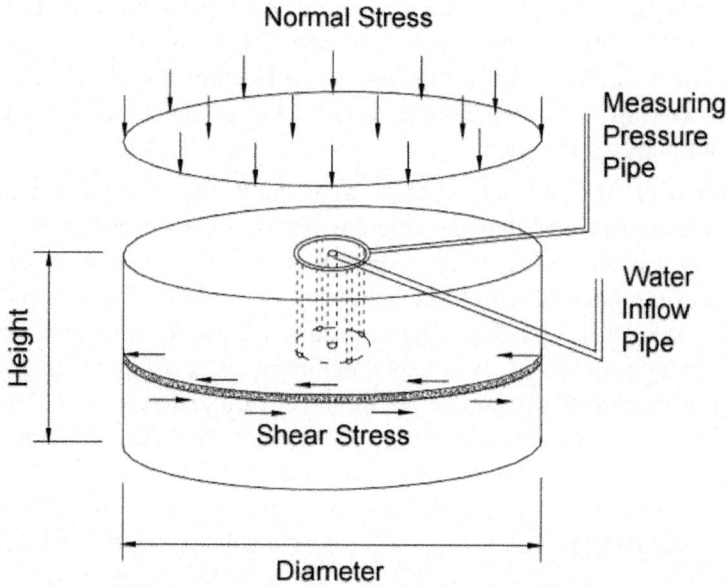

Figure 4. Sketch of the marble specimen for shear-flow coupling test.

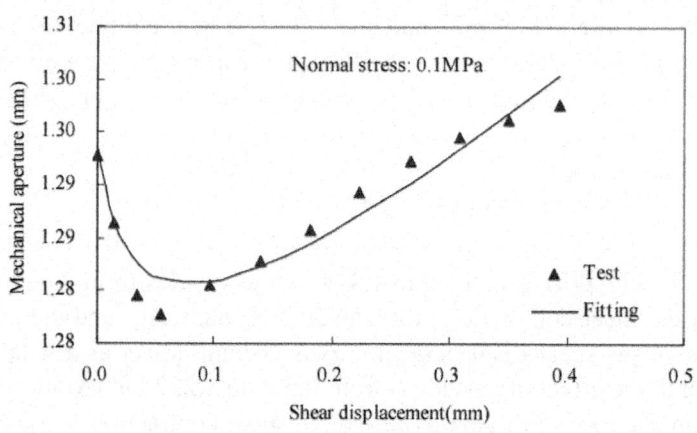

(a)

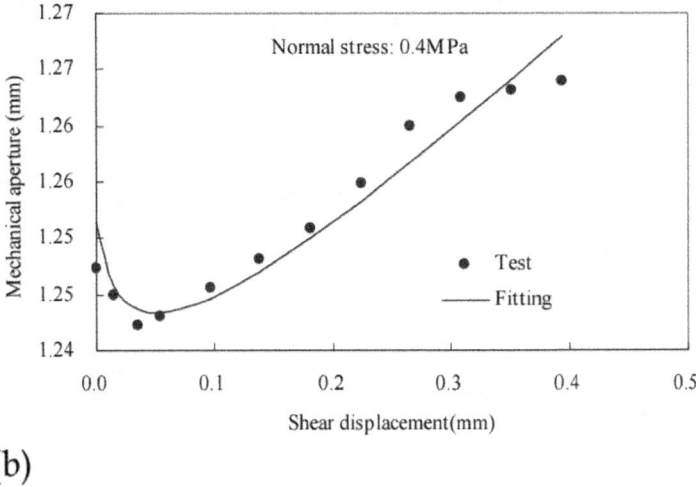

(b)

Figure 5. Mechanical aperture versus shear displacement curve under constant normal stress: (a) Normal stress: 0.1 MPa and (b) Normal stress: 0.4 MPa.

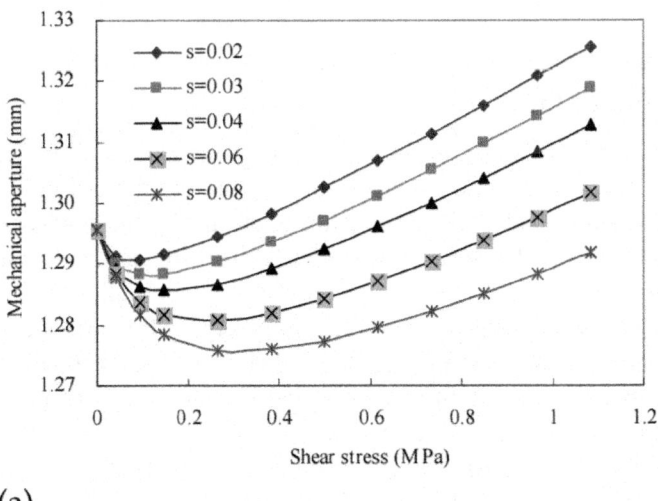

(a)

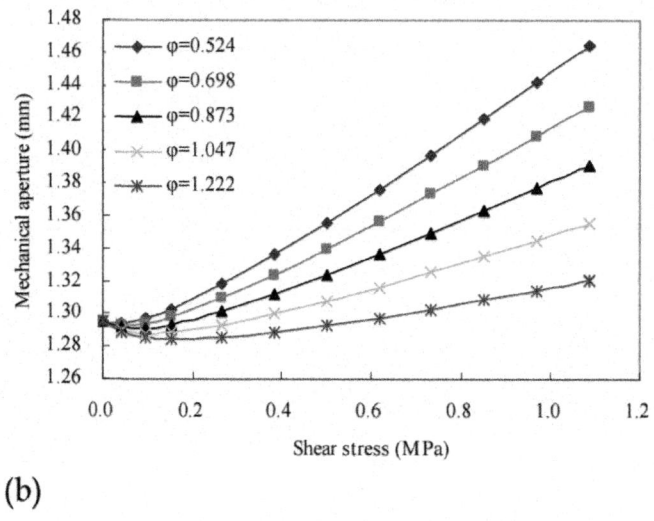

(b)

Figure 6. The sensitivity of s and φ on the behavior of the fracture: (a) φ=1.324 and (b) s=0.062.

STRAIN-DEPENDENT HYDRAULIC CONDUCTIVITY OF ROCK FRACTURES

In this section, we develop an elasto-plastic constitutive model for single hard rock fractures with consideration of nonlinear normal deformation and post-peak shear dilatancy, and then formulate the strain-dependent hydraulic conductivity for the fractures under dilated shear loading. Compared with the stress-dependent model presented in Section 2, one major difference is that the strain-dependent model is capable of describing the influence of postpeak mechanical response on the hydraulic properties of the fractures. This work is of paramount importance in the sense that the theoretical results are directly comparable with the experimental data of coupled shear-flow test, e.g. in Esaki et al. (1999). The straindependent hydraulic conductivity tensor can then be developed on this basis, which will be presented later in Section 5.

An Elasto-Plastic Constitutive Model for Rock Fractures

The underlying physical model considered is the same with the model plotted in Fig. 3, in which a fracture of hard rock is located at the mid-height of a specimen between two intact rock blocks. The height of the specimen is denoted by s, and the initial aperture of the fracture is b_0. When constant normal stress σ_n and increasing shear displacement δ are applied on the specimen, typical and idealized curves of shear displacement versus shear stress and shear displacement versus normal displacement (i.e. $\delta\sim\tau$ curve and $\delta\sim u$ curve) are plotted in Fig. 7. The shear stress increases linearly with the shear displacement (linked by the initial shear stiffness of the fracture, k_{s0}) until the shear stress approaches the peak, τ_p, which is then followed by a shear softening process in which the shear stress decreases to a residual level at a decreasing gradient with increasing shear displacement. For the purpose of deriving the hydraulic property of the fracture in post-peak loading section, however, an elastic-perfectly plastic $\delta\sim\tau$ relationship can be assumed, as shown in Fig. 7(a)

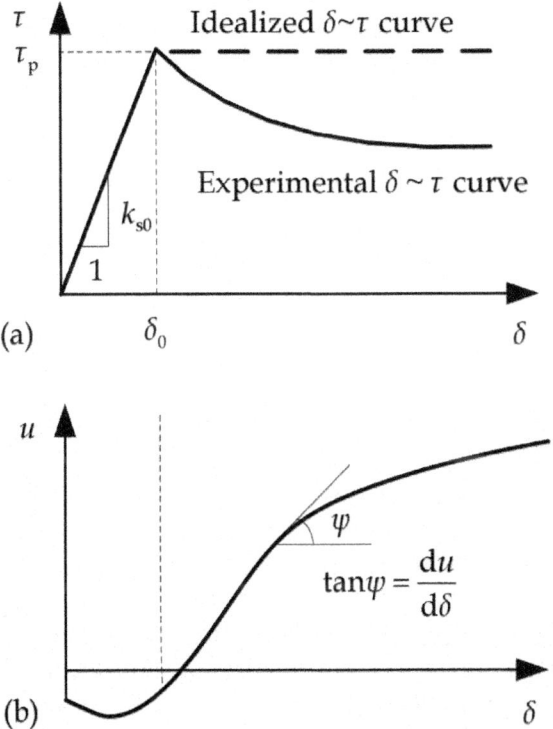

Figure 7. Typical and idealized curves of shear displacement versus shear stress and shear displacement versus normal displacement of a fracture subjected to normal and shear loads.

The deformation response of a rock fracture subjected to normal and shear loadings includes two components: one is the nonlinear closure of the fracture due to normal compression, and the other is the opening of the fracture due to shear dilation. Experimental results in Esaki et al. (1999) show that in the shearing process under constant normal loading, dilatancy will start when the shear stress approaches the peak and it increases at a decreasing gradient with increasing shear displacement, as illustrated in Fig. 7(b). As a result, the aperture of the fracture and then the hydraulic conductivity vary with increasing shear displacement.

Therefore, we may consider that shear dilatancy as well as the change in hydraulic conductivity accompanies normal and plastic shear deformations of the fracture. To deduce the hydraulic conductivity of the fracture with an averaging method, which will be further used later for deriving the hydraulic conductivity tensor for fractured rocks, we view the specimen with fracture as an equivalent continuous medium, i.e. the hydromechanical properties of the fracture are averaged into the whole specimen. As can be seen later, such a treatment does not affect our final solution to a single fracture, but it renders valid the small strain assumption on the fractures in the presence of post-sliding plasticity.

For a one-dimensional problem with a single rock fracture, the elasto-plastic constitutive model can be represented in the following forms:

$$\gamma_p = \gamma - \gamma_e = \frac{\delta}{s} - \frac{\delta_0}{s} = \frac{\delta}{s} - \frac{\tau_p}{sk_{s0}} \tag{23}$$

$$\varepsilon_n = \frac{\sigma_n'}{sk_n} + \int \tan\psi d\gamma_p \tag{24}$$

where γ, γ_e and γ_p are the total shear strain, the elastic shear strain and the plastic shear strain of the fracture, respectively; ε_n is the normal strain of the fracture; τ_p is the peak shear stress of the fracture under effective normal stress σ_n'; k_n and k_{s0} are, respectively, the normal stiffness and the initial shear stiffness of the fracture; δ_0 is the maximum elastic shear displacement upon shear yielding, with $\delta_0 = \tau_p/k_{s0}$, as shown in Fig. 7(a); and ψ is the mobilized dilatancy angle of the fracture. Note that in Eq. (24), the first term on the right hand side denotes the nonlinear closure of the fracture subjected to effective normal stress σ_n', while the second term denotes the opening of the fracture due to shear dilatancy.

Existing studies have indicated that shear dilatancy is highly dependent on the plasticity already experienced by the fractures and normal stress, and non-negligibly dependent on scale (Barton & Bandis, 1982; Yuan & Harrison, 2004; Alejano & Alonso, 2005). The decaying process of the dilatancy angle in line with plasticity can be described by the following negative exponential expression through the plastic shear strain, γ_p, or indirectly through the plastic shear displacement, δ, on the basis of Eq. (23):

$$\psi = \psi_{peak}\exp\left[-r(\delta - \delta_0)\right] \tag{25}$$

where r is a parameter for modelling the rate of decay that ψ undergoes as the plastic shear strain evolves. If r=0, then a constant dilatancy angle is recovered. As r→∞, the dilatancy angle quickly decays to zero. ψ_{peak} is the peak dilatancy angle of the fracture in the form of (Barton & Bandis, 1982)

$$\psi_{peak} = JRC \cdot \log_{10} \frac{JCS}{-\sigma'_n}$$

(26)

where JRC and JCS are the roughness coefficient and the wall compressive strength of fractures, respectively, and the actual values of them should be scale-corrected (Barton & Bandis, 1982). Thus, the dependencies of fracture dilatancy on plasticity, normal stress and scale are established through Eqs. (25) and (26).

Note that Eq. (25) shares the same shape with the asperity angle proposed for the description of shear dilatancy and surface degradation (Plesha, 1987), but the latter is represented as a function of the plastic tangential work. With the assumption of elasticperfectly plasticity, they are fully equivalent for monotonic loading (Jing et al., 1993). Cyclic loading is not a concern in this simple model, but when cyclic loading is involved, another independent function can be associated to the reverse loading that starts from the original point, just as the suggestion given in Plesha (1987) for asperity angles in two opposite directions, in order to satisfy the thermodynamic restriction condition presented in Jing et al. (1993).

Using the Mohr-Coulomb criteria, the peak shear stress τ_p of the fracture under effective normal stress σ'_n satisfies

$$\tau_p = -\sigma'_n \tan\varphi + c$$

(27)

where φ and c are the frictional angle and the cohesion of the fracture.

Differentiating Eq. (23) yields

$$d\gamma_p = d\gamma = \frac{1}{s} d\delta$$

(28)

Combining Eqs. (24) and (28) results in

$$\Delta b \approx s\varepsilon_n = \frac{\sigma'_n}{k_n} + \int_{\delta_0}^{\delta} \tan\psi(\delta)d\delta$$

(29)

An interesting phenomenon in Eq. (29) is, as described before, the change in the aperture of the fracture, Δb, is irrelevant to the height of the specimen, s. To conveniently use this formulation, two remedies can be further made:

First, suppose that the hyperbolic variation of k_n with the increase of aperture can be considered in the following (Huang et al., 2002):

$$k_n = \frac{-\sigma_n' + b_0 k_{n0}}{b_0}$$

(30)

where k_{n0} is the initial normal stiffness of the fracture.

Second, by employing the Taylor series expansion (truncated at the third order term), $\tan\psi$ can be adequately approximated by $\psi + \psi^3/3$ in radians for a rather large ψ_{peak}, e.g. 30°.

From Eq. (29) and the above two remedies, we have

$$\Delta b = \chi b_0$$

(31)

$$b = b_0 + \Delta b = (1 + \chi) b_0$$

(32)

with the parameter, χ, in the following form

$$\chi = \frac{\sigma_n'}{-\sigma_n' + b_0 k_{n0}} + \frac{1}{b_0} \left\{ \frac{\psi_{peak}}{r} \left[1 - e^{-r(\delta - \delta_0)} \right] + \frac{\psi_{peak}^3}{9r} \left[1 - e^{-3r(\delta - \delta_0)} \right] \right\}$$

(33)

Strain-Dependent Hydraulic Conductivity for Rock Fractures

Rewrite from Eq. (22) the initial hydraulic conductivity of the fracture, k_0, in the following form:

$$k_0 = \varsigma \frac{g b_0^2}{\nu}$$

(34)

Then, the hydraulic conductivity of the fracture under effective normal stress σ_n' and shear displacement δ can be described by

$$k = \varsigma \frac{g b^2}{\nu} = k_0 (1 + \chi)^2$$

(35)

Hence, a theoretical model of the hydraulic conductivity for a single rock fracture is finally formulated, which is totally determined by the effective normal stress σ_n' and the shear displacement δ, as well as a set of parameters characterizing the behavior of the fracture (i.e. b_0, ς, k_{n0}, k_{s0}, φ, c, JRC, JCS and r, which all can be deduced or back-calculated from experimental data). Note that by Eqs. (35) and (33), the proposed hydraulic conductivity model for rock fractures subjected to normal and shear loadings with mobilized dilatancy behavior depends in form on the plastic shear displacement, but from Eq. (23), one observes that the model depends indirectly on the plastic shear strain. Thus, we classify the established model into the stain-dependent hydraulic conductivity model.

Validation of the Proposed Model

Esaki et al. (1999) systematically investigated the coupled effect of shear deformation and dilatancy on hydraulic conductivity of rock fractures by developing a new laboratory technique for coupled shear-flow tests of rock fractures. In this section, we validate the theory proposed in Section 3.2 using the experimental data reported in Esaki et al. (1999). For this purpose, we first briefly introduce the experiments, and then predict our analytical results through Eqs. (31) and (35) by directly comparing with the experimental data.

The Coupled Shear-Flow Tests

The coupled shear-flow tests were conducted with an artificially created granite fracture sample under various constant normal loads and up to a residual shear displacement of 20 mm (Esaki et al., 1999). The underlying specimen for coupled shear-flow tests is sketched in Fig. 3, with its size of 120 mm in length, 100 mm in width and 80 mm in height. The initial aperture of the created fracture, b_0, is about 0.15 mm. The value of JRC is 9, and the value of JCS is 162 MPa, respectively.

The coupled shear-flow tests were conducted by first applying a prescribed normal stress ranging between 1 MPa and 20 MPa and then applying shear displacement in steps at a rate of 0.1 mm/s until a maximum shear displacement of 20 mm was reached. During tests, steady-state fluid flow rate, shear loading and dilatancy were all continuously recorded. The hydraulic aperture and conductivity were back-calculated by applying the cubic law, with the flow equations solved by using a finite difference method.

Determination of the Parameters for the Proposed Model

Some of the experimental values of the mechanical parameters of the fracture specimen during the coupled shear-flow tests are listed in Table 2 (taken from Table 1 in Esaki et al. (1999)). Using the data as listed in Table 2, we plot the peak shear stress versus normal stress curve in Fig. 8, which can be fitted by a linear equation $\tau_p = 1.058\sigma_n + 0.993$ with a high correlation coefficient of 0.9999.

Therefore, the shear strength of the specimen can be derived as $\varphi=46.6°$ and $c=0.99$ MPa, respectively.

Table 2. Mechanical parameters of the artificial fracture (After Esaki et al. (1999))

σ_n (MPa)	τ_p (MPa)	k_{s0} (MPa/mm)
1	2.06	3.37
5	6.16	10.65
10	11.74	11.97
20	22.10	17.97

The initial normal stiffness of the fracture of the specimen, k_{n0}, has to be estimated from the recorded initial normal displacement with zero shear displacement under different normal stresses. From the data plotted in Fig. 9 (which is taken from Fig. 7b in Esaki et al. (1999)), k_{n0} can be estimated as $k_{n0}=100$ MPa/mm by considering the possible deformation of the intact rock under high normal stresses. It is to be noted that in the remainder of this section, the hard intact rock deformation of the small specimen is neglected, meaning that the normal displacement of the specimen mainly occurs in the fracture of the specimen and it is approximately equal to the increment of the mechanical aperture of the fracture.

Theoretically, the decay coefficient of the fracture dilatancy angle, r, can be directly measured from the normal displacement versus shear displacement curves as plotted in Fig. 9. A better alternative, however, is to fit the experimental curves using Eq. (31) such that the least square error is minimized. By this approach, we obtain that r=0.13 with a correlation coefficient of 0.9538.

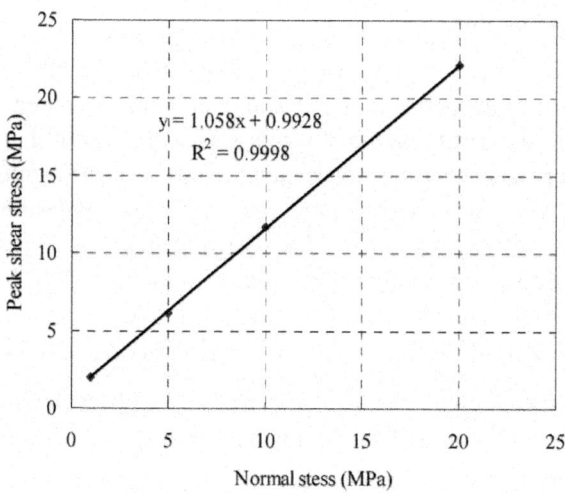

Figure 8. Peak shear stress versus normal stress curve of the fracture.

To obtain the dimensionless constant, ς, in Eq. (35) that relates the mechanical aperture to the hydraulic conductivity of the fracture under testing, further efforts are needed. A simple approach is to back-calculate ς directly using Eq. (34) with initial hydraulic conductivity, k0. But similarly, the better alternative is to fit the hydraulic conductivity versus shear displacement curves, as plotted in Fig. 11 (which is taken from Fig. 7c-f in Esaki et al. (1999)), using Eq. (35) such that the least square error is minimized. With such a method, we obtain that $\varsigma=0.00875$. This means that the mechanical aperture, b, and the hydraulic aperture, b^*, are linked with $b^*=0.324b$, which is very close to the experimental result shown in Fig. 8 in Esaki et al. (1999).

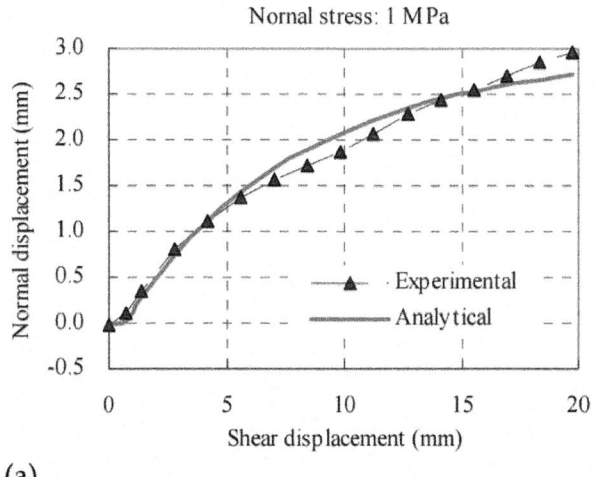

(a)

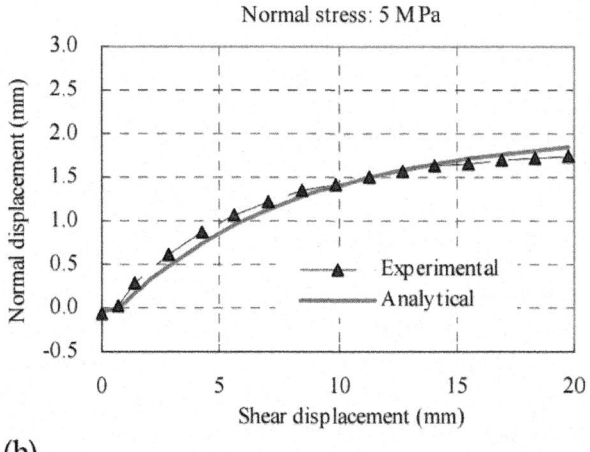

(b)

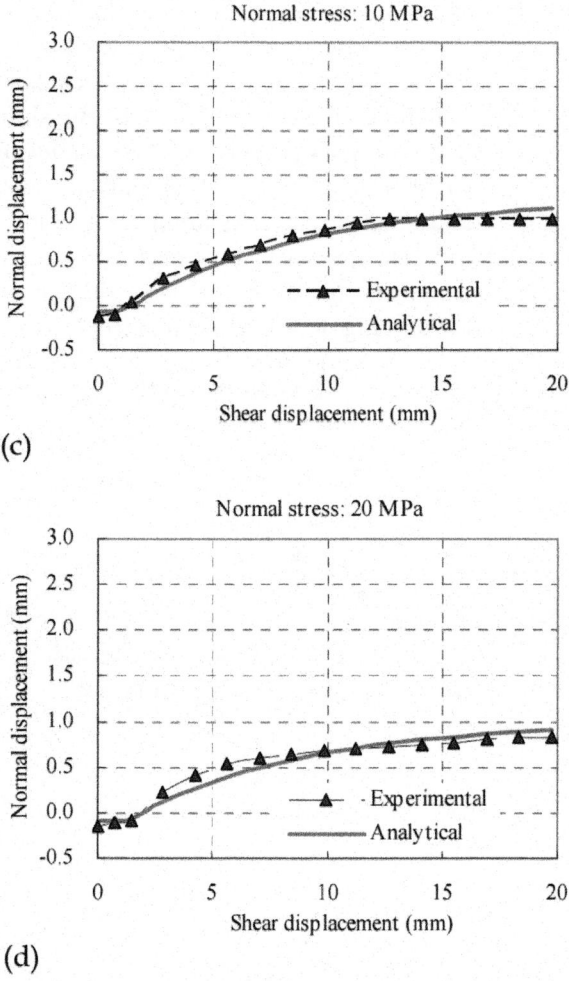

(c)

(d)

Figure 9. Comparison of the fracture aperture analytically predicted by Eq. (31) with that measured in coupled shear-flow tests.

Validation of the Proposed Theory

With the necessary parameters obtained in Section 3.3.2, we are now ready to compare the proposed model in Eqs. (31) and (35) with the experimental data presented in Esaki et al. (1999). Note that although the experimental data are available for one cycle of forward and reverse shearing, only the results for the forward shearing part are considered. The reverse shearing process, however, can be similarly modelled. Fig. 9 depicts the relations between the mechanical aperture and shear displacement that were measured from the coupled shear-flow tests presented in Esaki et al. (1999) and predicted by using the proposed model given in Eq. (31) under different normal stresses applied during the testing. It can be observed from Fig. 9 that our proposed analytical model is able to describe the shear dilatancy behavior of a real fracture under wide range of normal stresses between 1 MPa and 20 MPa by feeding appropriate parameters. Even the fracture aperture increases by one order of magnitude due to shear dilation, the analytical model still fitted the experimental results well. For practical uses, the slight discrepancies between the analytical results and the experimental data are negligible and the proposed model is accurate enough to characterize the significant dilatancy behavior of a real fracture.

This performance is largely attributed to the dilatancy model introduced through Eqs. (25) and (26). The dilatancy angles of the fracture evolving with the plastic shear displacement under different normal stresses are illustrated in Fig. 10. The high dependencies of the dilatancy angle of the fracture on normal stress and plasticity are clearly demonstrated in the curves. The peak dilatancy angle, which can be rather accurately modelled by Barton's peak dilatancy relation (Barton & Bandis, 1982), decreases logarithmically with the increase of the applied normal stress. For normal stresses of 1 MPa, 5 MPa, 10 MPa and 20 MPa, the peak dilatancy angles are 19.9°, 13.6°, 10.9° and 8.2°, respectively. On the other hand, the dilatancy angle undergoes negative exponential decay with increasing plastic shear displacement, a process related to surface degradation of rough fractures.

Fig. 11 shows the hydraulic conductivity versus shear displacement relations that were back-calculated from fluid flow results using the finite difference method from the coupled shear-flow tests presented in Esaki et al. (1999) and that are predicted by the proposed model given in Eq. (35) under different normal stresses during testing. As shown in the semi-logarithmic graphs in Fig. 11, the proposed analytical model can well predict the evolution of hydraulic conductivity of the tested rock fracture, with the change in the magnitude of 2 orders, during coupled shear-flow tests under different normal stresses. The ratios of the predicted hydraulic conductivities to the corresponding experimental results all fall in between 0.3 and 3.0, indicating that they are rather close in orders of magnitude and the predicted results are suitable for practical use.

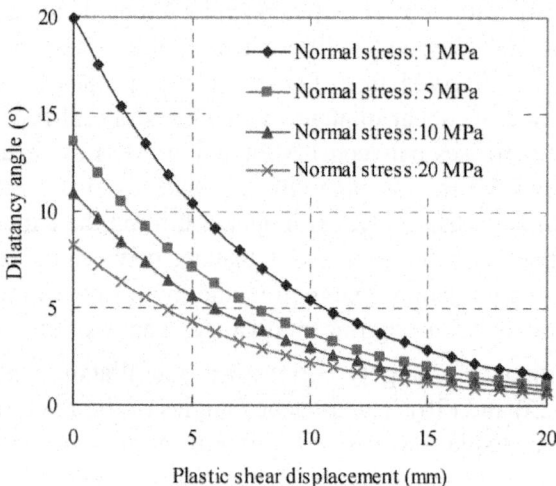

Figure 10. Dilatancy angles of the fracture evolving with the plastic shear displacement under different normal stresses.

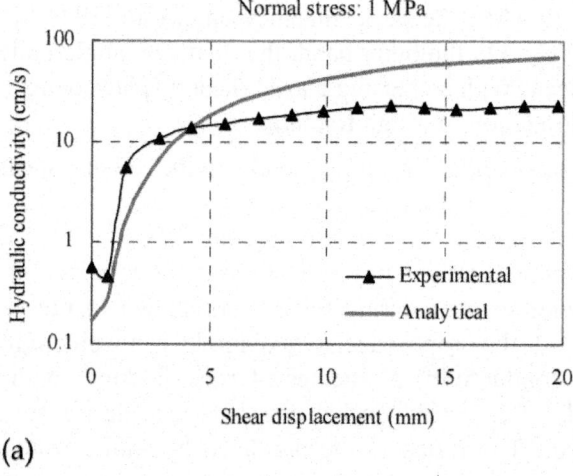

(a)

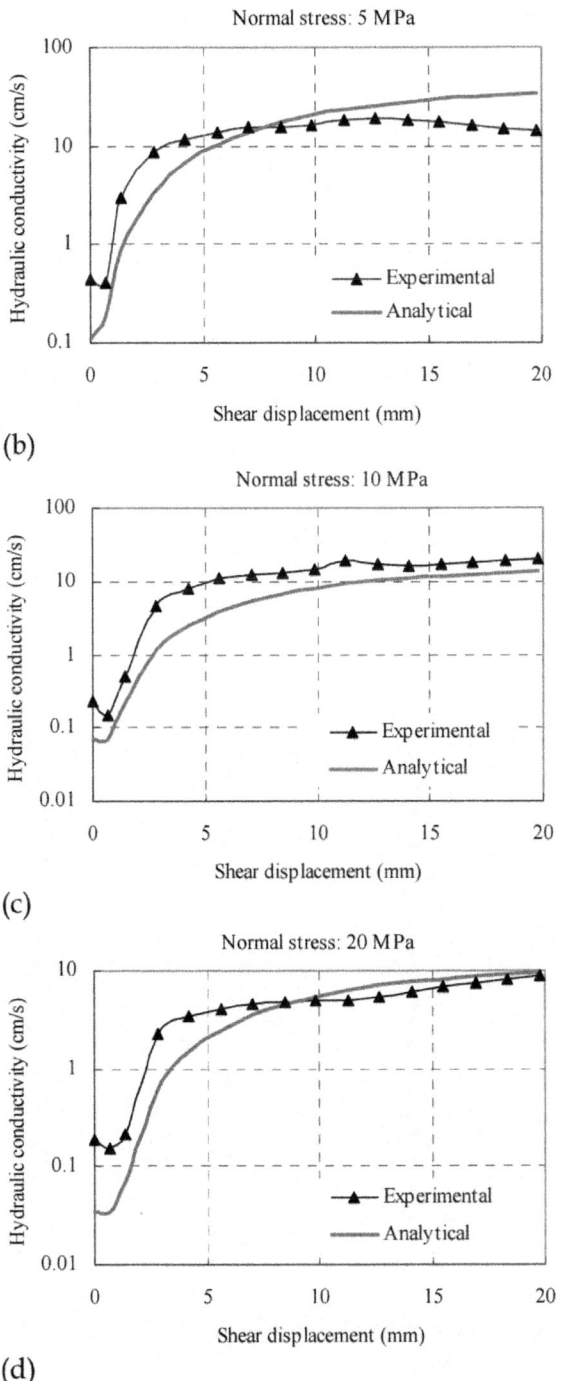

Figure 11. Comparison of the hydraulic conductivity analytically predicted by

Eq. (35) with that calculated from coupled shear-flow tests with finite difference method.

STRESS-DEPENDENT HYDRAULIC CONDUCTIVITY TENSOR OF FRACTURED ROCKS

When the response of each fracture under normal and shear loading is understood (see Section 2), the remaining problem is how to formulate the hydraulic conductivity for fractured rock mass based on the geometry of the underlying fracture network. Fig. 12 depicts a two-dimensional fracture network (taken after Min et al. (2004)) in a biaxial stress field. As shown in Fig. 12, each fracture plays a role in the hydraulic conductivity of the rock mass, and its contribution primarily depends on its stress state, its occurrence, as well as its connectivity with other fractures. Also shown in Fig. 12 is the scale effect of the rock mass on hydraulic properties. When the size of the rock mass is small, only a few number of fractures are included and heterogeneity of the hydraulic conductivity of the rock mass may dominate. As the population of factures grows with the increasing size, an upscaling scheme may be available to derive a representative hydraulic conductivity tensor for the rock mass at the macroscopic scale.

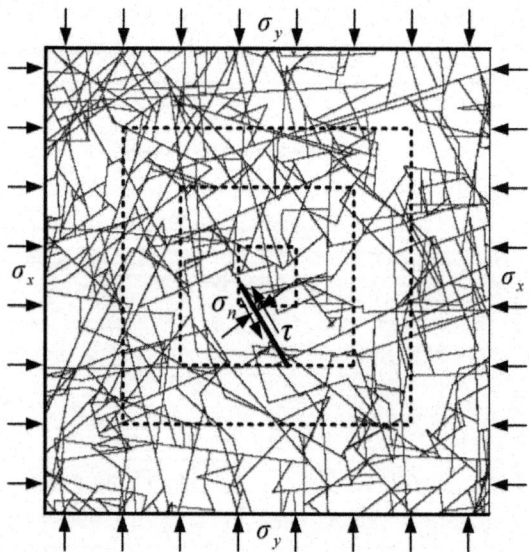

Figure 12. A fracture network (taken after Min et al. (2004)) in biaxial stress field and the scale effect of the rock mass.

Based on the above observations, in this section, we formulate an equivalent hydraulic conductivity tensor for fractured rock mass based on the superposition principle of liquid dissipation energy, in which the concept of REV is integrated and the applicability of an equivalent continuum approach is able to be validated.

Computational Model

Without loss of generality, the global coordinate system $X_1X_2X_3$ is established in such a way that its X_1-axis points towards the East, X_2-axis toward the North and X_3-axis vertically upward. A local coordinate system $x_1^f x_2^f x_3^f$ is associated with the *fth* set of fractures such that the x_1^f-axis is along the main dip direction, the x_2^f-axis is in the strike, and the x_3^f-axis is normal to the fractures, as shown in Fig. 13.

In order to formulate the stress-dependent hydraulic conductivity tensor for fractured rock masses using the aforementioned elastic constitutive model for rock fractures, the following assumptions, similar to Oda (1986), are made in this section:

- A cube of volume, V_p, is considered as the flow region of interest, which is cut by n sets of fractures. The orientation of each set of fractures is indicated by a mean azimuth angle β and a mean dip angleα. Other geometrical statistics of the fractures are assumed to be available through field measurements or empirical estimations.

- Even though the geometry of real fractures is complex, generally it can be simplified as a thin interfacial layer with radius r and aperture b*.

- The rock mass is regarded as an equivalent continuum medium, which means the representative elementary volume (REV) exists in the rock mass and its size is smaller than or equal to V_p.

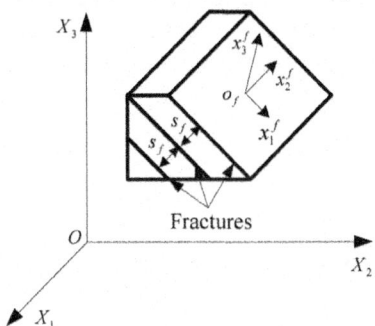

Figure 13. Coordinate systems.

Stress-Dependent Hydraulic Conductivity Tensor

Fluid flow through the equivalent continuum media can be described by the generalized 3- D Darcy's law as follows:

$$v = \mathbf{KJ} \tag{36}$$

where v denotes the vector of flow velocities, J denotes the vector of hydraulic gradients, and K is the hydraulic conductivity tensor for the rock mass.

For steady-state seepage flow, the dissipation energy density, $e(X_1, X_2, X_3)$, of fluid flow through the media can be represented as (Indelman & Dagan, 1993):

$$e = \frac{1}{2} \mathbf{J}^{T} \mathbf{KJ} \tag{37}$$

Hence, the total flow dissipation energy, E, in the rock mass V_p can be calculated by performing an integration throughout the whole flow domain:

$$E = \int_{V_p} e d\Omega = \frac{1}{2} \int_{V_p} \mathbf{J}^{T} \mathbf{KJ} d\Omega \tag{38}$$

If REV does exist in the rock mass and its size is smaller than or equal to V_p, by defining $\bar{\mathbf{J}}$ to be the vector of the average hydraulic gradient within V_p and $\bar{\mathbf{K}}$ to be the average hydraulic conductivity tensor, Eq. (38) can be reduced to:

$$E = \frac{1}{2} \bar{\mathbf{J}}^{T} \bar{\mathbf{K}} \bar{\mathbf{J}} V_p \tag{39}$$

Suppose that the volume density of the ith set of fractures is J_{vi}. The number of this set of fractures can be estimated by $m_i = J_{vi} V_p$.

For permeable rock matrix, the flow dissipation energy shown in Eq. (39) consists of two components, i.e., the flow dissipation energy through rock matrix, E_r, and the flow dissipation energy through crack network, E_c:

$$E = E_r + E_c \tag{40}$$

E_r can be represented as:

$$E_r = \frac{1}{2} \bar{\mathbf{J}}^{T} \bar{\mathbf{K}}_r \bar{\mathbf{J}} V_p \tag{41}$$

where $\bar{\mathbf{K}}_r$ denotes the hydraulic conductivity tensor for rock matrix. If rock matrix is impermeable, all elements in $\bar{\mathbf{K}}_r$ vanish. To estimate E_c, we introduce a weight coefficient W_{ij} to describe the effect of the connectivity of the fracture network on fluid flow:

$$W_{ij} = \xi_{ij} / \bar{\xi}_i \tag{42}$$

where ξ_{ij} is a stochastic variable denoting the number of fractures intersected by the jth fracture belonging to the ith set; and $\bar{\xi}_i$ denotes the maximum number of fractures cut by the ith set of fractures. Obviously, $0 \le W_{ij} \le 1$ and when $\xi_{ij} = 0$, $W_{ij} = 0$. This implies that an entirely isolated fracture which does not intersect any other fracture effectively contributes nothing to the hydraulic conductivity of the total rock mass.

For the jth fracture belonging to the ith set, a void volume equal to $\pi r_{ij}^2 b_{ij}^*$ is associated with it. Then, the flow dissipation energy through it is described as:

$$E_{cij} = W_{ij} e_{ij} \pi r_{ij}^2 b_{ij}^* \tag{43}$$

where e_{ij} is shown as follows:

$$e_{ij} = \frac{1}{2} k_{ij} \bar{J}_{ci}^T \bar{J}_{ci} \tag{44}$$

where k_{ij} denotes the hydraulic conductivity of the jth fracture of the ith set, which can be calculated by the stress-dependent hydraulic conductivity model, Eq. (21).

$\bar{J}_{ci}$ denotes the hydraulic gradient within the ith set of fractures:

$$\bar{J}_{ci} = (\delta - \mathbf{n}_i \otimes \mathbf{n}_i)\bar{J} \tag{45}$$

where δ is the Kronecker delta tensor, and $\mathbf{n}_i$ denotes the unit vector normal to the ith set of fractures, with its components $n_1 = \sin\alpha\sin\beta$, $n_2 = \sin\alpha\cos\beta$, and $n_3 = \cos\alpha$.

Thus, E_c can be represented as

$$E_c = \frac{8\pi}{12\nu} \sum_{i=1}^{n} \sum_{j=1}^{m_i} W_{ij} r_{ij}^2 b_{ij}^{*3} \bar{J}^T (\delta - \mathbf{n}_i \otimes \mathbf{n}_i)\bar{J} \tag{46}$$

From Eqs. (39)-(41), (46) and (20), it can be referred that

$$\bar{K} = \bar{K}_r + \frac{8\pi}{12\nu V_p} \sum_{i=1}^{n} \sum_{j=1}^{m_i} W_{ij} f^3(\beta_{ij}) r_{ij}^2 b_{0ij}^3 (\delta - \mathbf{n}_i \otimes \mathbf{n}_i) \tag{47}$$

In Eq. (47), n is determined by the orientation of the fractures, which reflects the effect of the orientation of the fractures on the fluid flow. r and b_0 represent the size or the scale of the fractures; they retrain the fluid flow through the fractures from their developing magnitude. W is a parameter introduced to show the impact of the connectivity of the fracture network on fluid flow. Finally, $f(\beta)$ is a function used to demonstrate the coupling effect between fluid flow and stress state.

The hydraulic tensor for fractured rock masses given in Eq. (47) is related to the volume of the flow region, V_p, which exactly shows the size effect of the hydraulic properties. Intuitively, the smaller the V_p size is, the less number of fractures is contained within the volume, and thus the poorer the representative of the computed hydraulic conductivity tensor. On the other hand, when V_p is increased up to a certain value, the fractures involved in the cubic volume are dense enough and the hydraulic conductivity tensor for the rock mass does not vary with the size of the volume. This V_p size is exactly the representative elementary volume, REV, of the flow region. The V_p size of the flow region is required to be larger than REV for estimating the hydraulic conductivity tensor for the fractured rock mass. Otherwise, treating the fractured rock mass as an equivalent continuum medium is not appropriate, and the discrete fracture flow approach is preferable.

Comparison with Snow's and Oda's Models

Now we make a comparison between the formulation of the hydraulic conductivity tensor presented in Eq. (47) and the formulation given by Snow (1969) as well as the formulation given by Oda (1986). The Snow's formulation is as follows:

$$\mathbf{K} = \frac{g}{12v} \sum_{i=1}^{n} \frac{b_i^3}{s_i} (\delta - \mathbf{n}_i \otimes \mathbf{n}_i)$$

(48)

where s_i is the average spacing of the ith set of fractures. If we neglect the hydraulic conductivity of the rock matrix and the connectivity of the factures, and define

$$b_i = \frac{1}{m_i} \sum_{j=1}^{m_i} f(\beta_{ij}) b_{0ij} \quad \text{and} \quad s_i^{-1} = \frac{\pi}{V_p} \sum_{j=1}^{m_i} r_{ij}^2$$

(49)

Then, the formulation presented in Eq. (47) is totally equivalent to Snow's formulation, Eq. (48).

On the other hand, the Oda's formulation is described by

$$\mathbf{K} = \varsigma(P_{kk}\boldsymbol{\delta} - \mathbf{P})$$

(50)

where P is the fracture geometry tensor, with $P_{kk} = P_{11} + P_{22} + P_{33}$.

$$\mathbf{P} = \pi\rho\int_0^\infty \int_0^\infty \int_\Omega r^2 b^3 \mathbf{n} \otimes \mathbf{n}E(n,r,b)\mathrm{d}\Omega \mathrm{d}r\mathrm{d}b$$

(51)

where E(n, r, b) is a probability density function of the geometry of the fractures, ρ is the number of fracture centers per unit of volume, with $\rho = m_v/V_p$, $m_v = \sum m_i$, and ς is the dimensionless scalar adopted to penalize the permeability of real fractures with roughness and asperities. Assuming that a statistically valid REV exists and being aware that the fracture orientation is a discrete event, the fracture geometry tensor may be empirically constructed by the following direct summation

$$\mathbf{P} = \frac{\pi}{V_p}\sum_{i=1}^{m_v} r_i^2 b_i^3 \mathbf{n}_i \otimes \mathbf{n}_i$$

(52)

Following a similar deduction, it can be inferred that all these three formulations are equivalent not only in form but also in function, though they are derived from different approaches and different assumptions. The formulation presented in Eq. (47) can be directly obtained from Snow's formulation by considering the connectivity and roughness of the fractures and integrating the aperture changes under engineering disturbance. The discretized form of the Oda's formulation is much closer to the current formulation, and the latter can also be directly achieved from the former by considering the connectivity of the fracture network. However, the proposed method for formulating an equivalent hydraulic conductivity tensor for complex rock mass based on the superposition principle of liquid dissipation energy is a widely applicable approach not only to equivalent continuum but also to discrete medium.

A Numerical Example: Hydraulic Conductivity of the Rock Mass in the Laxiwa Hydropower Project

In order to validate the theoretical model presented in Section 4.2, we investigated the hydraulic conductivity of a fractured rock mass at the construction site of the Laxiwa Hydropower Project, the second largest hydropower project on the upstream of the Yellow River. The selected construction site for a double curvature arch dam is a V-shaped valley formed by granite rocks, as shown in Fig. 14. The dam height is 250 m, the top elevation of the dam is 2460 m, the reservoir storage capacity is 1.06 billion m³ and the total installed capacity is 4200 MW.

A typical section of the Laxiwa dam site is illustrated in Fig. 15. Besides faults, four sets of critically oriented fractures are developed in the rock mass at the construction site. The geological characteristics of the fractures are described by spacing, trace length, aperture, azimuth, dip angle, the joint roughness coefficient, JRC, of the fractures as well as the connectivity of the fracture network (i.e., the number of fractures intersected by one fracture). According to site investigation, the statistics (i.e., the averages and the mean squared deviations, as well as the distribution of the characteristics) of the fractured rock mass on the right bank of the valley are listed in Table 3.

Figure 14. Site photograph of the Laxiwa valley.

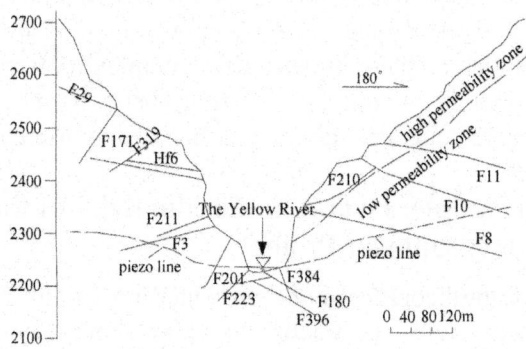

Figure 15. A typical section of the Laxiwa dam site.

Table 3. Characteristic variables of the fractured rock mass[*]

Set	Spacing (m)	Length (m)		Aperture (mm)		Azimuth (°)		Dip (°)		Connectivity	
		avg.	dev.	avg.	dev.	avg.	dev.	avg.	dev.	avg.	dev.
1	1.45	5	1.5	0.096	0.02	85.3	10	54.5	10	5	3
2	2.62	3	1.0	0.096	0.02	355.1	20	29.8	5	3	2
3	10.96	3	1.0	0.096	0.02	287.4	20	61.4	10	3	2
4	10.96	3	1.0	0.096	0.02	320.2	20	11.9	5	3	2
Distribution	logarithmic normal	negative exponential		Gama		normal		normal		normal	

[*]'avg.' denotes arithmetic mean of a variable,
'dev.' represents root mean squared deviation

At the construction site of the Laxiwa dam, a total number of 1450 single-hole packer tests were conducted to measure the hydraulic properties of the rock mass, with 113 packer tests for the shallow rock mass on the right bank in 0–80 m horizontal depth and 278 packer tests for the deeper rock mass. The measurements of the hydraulic conductivity range from 10^{-5} cm/s to 10^{-6} cm/s for the shallow rock mass and from 10^{-6} cm/s to 10^{-7} cm/s for the deeper rock mass, with in average 4.94×10^{-5} cm/s for the former and 3.80×10^{-6} cm/s for the latter, respectively (Liu, 1996). On the other hand, in-situ stress tests showed that the geostress in the base of the valley and in deep rock mass has a magnitude of 20–60 MPa, with the direction of the major principal stress pointing towards NNE. As a result of stress release, the release fractures are frequently developed and a high permeability zone of 0–80 m horizontal depth is formed in the bank slope, as shown in Fig. 15. The stress release fractures, however, become infrequent in deeper rock mass, and the measured hydraulic conductivity is generally 1–2 orders of magnitude smaller than the hydraulic conductivity of the rock mass in shallow depth away from the bank slope. Therefore, the hydraulic conductivity of the rock mass at the construction site of the Laxiwa arch dam is mainly controlled by the fracture network and the stress state.

Based on these statistics given in Table 3, fracture networks can be generated and calibrated for the rock mass at the construction site of the Laxiwa Hydropower Project using the MonteCarlo method by assuming that each fracture is a smooth, planar disc, with its center uniformly distributed in the simulated area. For each set of fractures, the geometrical parameters of any one are sampled by Monte-Carlo method until enough fractures are included in the simulated area. Then, a calibration procedure is invoked to check whether the generated model satisfies the distribution mode of the real fracture network. If doesn't, the fracture network will be regenerated until one matches the distribution mode. With the generated fracture network, the actual connectivity can be computed by spatial operation on the fractures.

But for calibrated fracture network, a more convenient approximate approach to determine the connectivity of the fracture network, as it is adopted here, is to directly produce ξ_{ij} in Eq. (42) with the Monte-Carlo method and the characteristics presented in Table 3, then W_{ij} is derived from Eq. (42) with $\bar{\xi}_i$, the maximum number of fractures cut by the ith set of fractures. Field measurements are used to estimate $\bar{\xi}_i$, with $\bar{\xi}_1=11$, $\bar{\xi}_2=8$ and $\bar{\xi}_3=\bar{\xi}_4=6$ for the four sets of fractures, respectively. Fig. 16 illustrates a simulated fracture network with size of 20×20×20 m.

On the basis of the fracture network generated above, we compute the hydraulic conductivity tensor for the simulated cubic volume of rock mass with size of 20×20×20 m using the method given by Snow (1969) and the method presented in Section 4.2, respectively. To show the coupling effect of stress/deformation on hydraulic properties, we consider two scenarios for examination. In the first scenario, we consider the fracture network located in the shallow depth away from the bank slope, where the impact of the in-situ stress is negligible. While in the second scenario, the fracture network is situated in larger depth, and a typical stress state with $\sigma_x=\sigma_z=10$ MPa and $\sigma_y=20$ MPa is associated with it. Based on laboratory test results, the shear modulus of the fractures is estimated as $\mu=2$ MPa, and then by taking the Poisson's ratio as $\nu=0.25$, the Lame's constant is derived with $\lambda=2$ MPa. The kinematic viscosity of underground water is set to be $v_w=1.14\times10^{-6}$ m2/s and the frictional angle-like parameter and the normal stress-like parameter are taken as $\varphi=0.4363$ and $s=\sigma_n/20$.

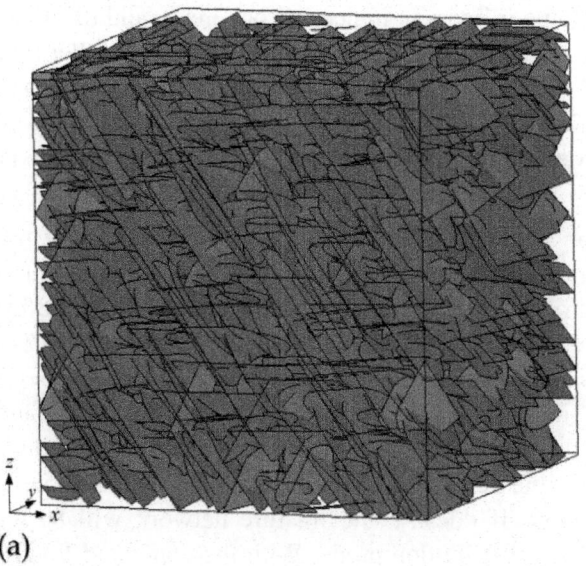

(a)

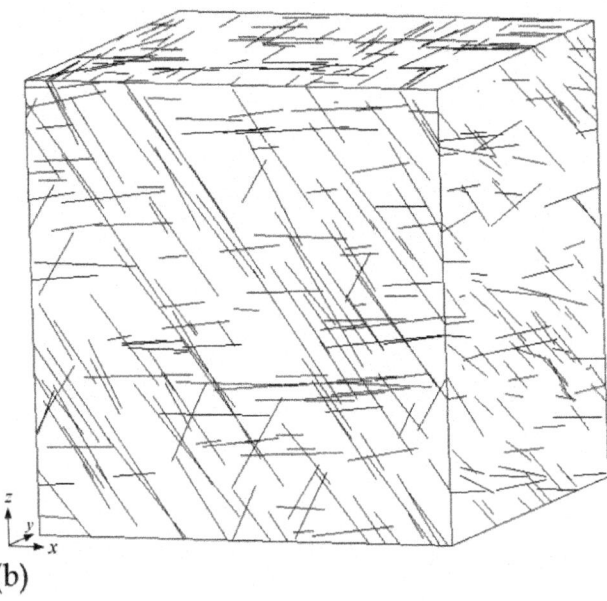

(b)

Figure 16. A three dimensional fracture network with size of 20×20×20 m generated by using the Monte-Carlo method for the rock mass in the Laxiwa Hydropower Project: (a) fracture network and (b) traces of the fractures on the surfaces of the simulated area.

The predicted hydraulic conductivity tensor for the examined rock mass is listed in Table 4. From Table 4, one observes that for shallow rock mass (where the effect of in-situ stress is not considered), the Snow's method and the method presented in Section 4.2 predict similar results and the predicted hydraulic conductivity is in the magnitude of 10^{-5} cm/s and close to in-situ hydraulic observations, but the anisotropy in hydraulic conductivity manifests due to non-uniform distribution of fractures. Compared with the hydraulic conductivity of the shallow rock mass, the predicted hydraulic conductivity for the rock mass in larger depth with the same fracture network decreases in 2 orders of magnitude due to the closure of the fractures applied by the in-situ stresses, but the anisotropic property of the hydraulic conductivity remains, which suggests that the occurrence of the fractures has a significant impact on permeability. Taking into consideration the applied stress level, the reduction of hydraulic conductivity in orders of magnitude is very close to the results achieved in Min et al. (2004) through a discrete element method, and generally agrees with the in-situ hydraulic observations.

Table 4. Predicted hydraulic conductivity tensor of the rock mass at the construction site of the Laxiwa dam (cm/s)

Snow's model (for shallow rock mass)		
4.78E–05	–4.76E–07	–1.71E–05
–4.76E–07	7.49E–05	–1.41E–05
–1.71E–05	–1.41E–05	4.08E–05
The proposed model (for shallow rock mass)		
1.93E–05	–1.75E–07	–6.39E–06
–1.75E–07	2.99E–05	–5.81E–06
–6.39E–06	–5.81E–06	1.64E–05
The proposed model (for deep rock mass)		
9.06E–08	–4.81E–09	–6.10E–08
–4.81E–09	1.85E–07	–1.92E–08
–6.10E–08	–1.92E–08	1.10E–07

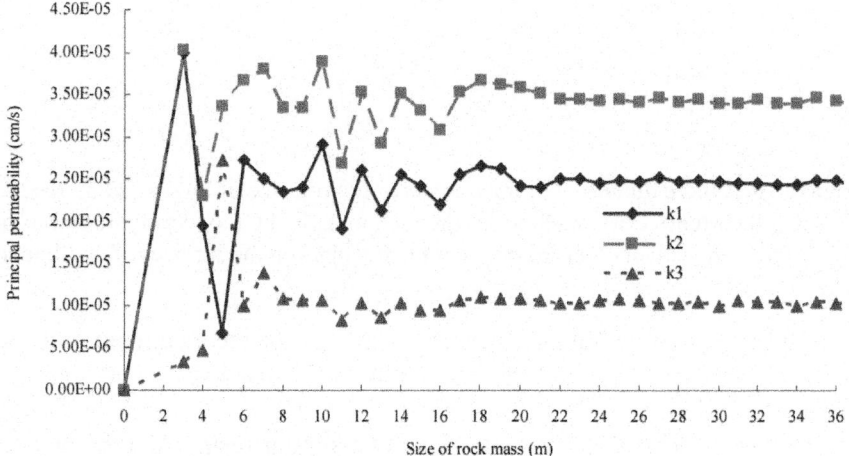

Figure 17. Hydraulic conductivity versus the volume size of the fractured rock mass.

Now, we take for example the rock mass in shallow depth to estimate the REV size of the rock mass. For this purpose, the scale of the rock mass is increased gradually from 3×3×3 m to 40×40×40 m with an increment of 1 m in each dimension. In each step, a fracture network with prescribed size is generated by using the Monte-Carlo method described above, and it is worth noting that this method is somewhat different from the method used by Min & Jing (2003) and Long et al. (1982). For each fracture network, the hydraulic conductivity tensor is calculated from Eq. (47) and then the principal hydraulic conductivities are further obtained from the hydraulic conductivity tensor. The relationship between the computed principal hydraulic conductivities and the sizes of the rock mass is illustrated in Fig. 17. As we can see from Fig. 17, when the block size of the rock mass is smaller than 18×18×18 m, the population of fractures is not dense enough and the principal hydraulic conductivities fluctuate dramatically. On the other hand, as the size scales up to about 20×20×20 m, the examined rock mass has included enough fractures and the computed principal hydraulic conductivities approach a rather steady state, with k_1, k_2, k_3 estimated to be $2.41×10^{-5}$ cm/s, $3.59×10^{-5}$ cm/s, $1.08×10^{-5}$ cm/s, respectively. This suggests that the REV does exist in the rock mass and the rock mass can be regarded as an equivalent continuum medium as long as its size is no less than, e.g., 20×20×20 m or 8000 m³.

STRAIN-DEPENDENT HYDRAULIC CONDUCTIVITY TENSOR OF FRACTURED ROCKS

On the basis of the strain-dependent model presented in Section 3 for rock fractures, this section formulates the strain-dependent hydraulic conductivity tensor for fractured rock masses cut by one or multiple sets of parallel fractures. The major difference between the model in this section and the stress-dependent model presented in Section 4 is that the former is capable of describing influence of the post-peak mechanical behaviors on the hydraulic properties of the rock masses, and is suited for modelling the coupled processes in rock masses at high stress level and in drastic engineering disturbance condition.

An Equivalent Elasto-Plastic Constitutive Model for Fractured Rocks

Consider a fractured rock mass cut by n sets of planar and parallel fractures of constant apertures with various orientations, scales and densities. The global response of the fractured rock mass under loading comes both from weak fractures and from stronger rock matrix. Based on this observation, an equivalent elasto-plastic constitutive model can be formulated by imposing assumptions on the interaction between fractures and rock matrix. The coordinate systems are defined in the same way with those defined in Section 4.1 (see Fig. 13). Denote the unit vector along X_i-axis of the global frame as e_i (i=1, 2, 3) and the unit vector along x_i^f-axis of the fth local frame as e_i^f (i=1, 2, 3). Then, a second order tensor, l f , can be defined for transforming physical quantities between the frames, with the components in the form of

$$l_{ij}^f = \mathbf{e}_i^f \cdot \mathbf{e}_j \qquad (53)$$

Regarding the fractured rock mass as a continuous medium at the macroscopic scale, it is rational to assume that the global strain increment of the fractured rock mass is composed of the strain increments of rock matrix and fractures (Pande & Xiong, 1982; Chen & Egger, 1999), i.e.

$$d\varepsilon = d\varepsilon^R + \sum_F d\varepsilon^F$$

(54)

where $d\varepsilon$, $d\varepsilon^R$ and $d\varepsilon^F$ are the total incremental strain tensor, the incremental strain tensor of rock matrix and the incremental strain tensor of fth set of fractures measured in the global coordinate system, respectively. Note that a variable with a superscript in upper case (i.e. R or F) means that it is measured in the $X_1X_2X_3$ system, while a variable with a superscript in lower case (i.e. f) is measured in $x_1^f x_2^f x_3^f$ system, respectively. Unless otherwise specified, the superscripts F and f are not summing indices.

On the other hand, traction continuity has to be ensured across the fracture interfaces. In the global coordinate system, this condition can be strictly represented by (Pande & Xiong, 1982; Chen & Egger, 1999)

$$d\sigma = d\sigma'^R = d\sigma'^F$$

(55)

where $d\sigma'$, $d\sigma'^R$ and $d\sigma'^F$ are the effective incremental stress tensor of the fractured rock mass, the effective incremental stress tensor of rock matrix and the effective incremental stress tensor of fth set of fractures, respectively. The effective stress tensor σ' is defined as

$$\sigma' = \sigma + \alpha p \delta$$

(56)

where σ is the total stress tensor (positive for tension), p is the pore water pressure (positive for compressive pressure), and α ($\alpha \leq 1$) is an effective stress parameter.

Combining the plastic potential flow theory and the consistency conditions of rock matrix and fractures, an equivalent elasto-plastic constitutive model can be derived from Eqs. (54) and (55):

$$d\varepsilon = S^{ep}:d\sigma'$$

(57)

with

$$S^{ep} = (C^{R,ep})^{-1} + \sum_F (C^{F,ep})^{-1}$$

(58)

where S^{ep} is the equivalent elasto-plastic compliance tensor of the fractured rock mass.

$C^{R,ep}$ in Eq. (58) is the elasto-plastic modulus tensor of rock matrix. Neglecting the degradation of rock strength in the volume close to fracture intersections, $C^{R,ep}$ can be written as

$$\mathbf{C}^{R,ep} = \mathbf{C}^R - \frac{\mathbf{C}^R : \dfrac{\partial Q_R}{\partial \boldsymbol{\sigma}'} \otimes \dfrac{\partial F_R}{\partial \boldsymbol{\sigma}'} : \mathbf{C}^R}{\dfrac{\partial F_R}{\partial \boldsymbol{\sigma}'} : \mathbf{C}^R : \dfrac{\partial Q_R}{\partial \boldsymbol{\sigma}'} + H_R}$$

(59)

where C^R is the fourth-order elastic modulus tensor of rock matrix, which can be represented in terms of the Lame's constants λ and μ:

$$C_{ijkl}^R = \lambda \delta ij \delta kl + \mu(\delta_{ik}\delta_{jl} + \delta_{il}\delta_{jk})$$

(60)

F_R, Q_R and H_R in Eq. (59) are the yield function, the plastic potential function and the hardening modulus of rock matrix, respectively. A non-associative flow rule with elasticperfectly plasticity (i.e. H_R=0) is adopted for better modeling dilatant behavior of rock matrix by virtue of, for example, the Druker-Prager criterion with its cone fully inscribed by the Mohr-Coulomb hexagon, defined by functions

$$F_R = aI_1' + \sqrt{J_2} - \kappa = 0$$

(61)

$$Q_R = \beta I_1' + \sqrt{J_2}$$

(62)

with

$$\alpha = \sin \varphi_R / \sqrt{3(3 + \sin^2 \varphi_R)}$$

(63)

$$\kappa = 3c_R \cos \varphi_R / \sqrt{3(3 + \sin^2 \varphi_R)}$$

(64)

$$\beta = \sin \psi_R / \sqrt{3(3 + \sin^2 \psi_R)}$$

(65)

where c_R and φ_R are the cohesion and the friction angle of rock matrix, respectively. I_1' and J_2 are the first invariant of the effective stress and the second invariant of the deviatoric stress of rock matrix, respectively. ψ_R is the mobilized dilatancy angle of rock matrix.

It should be noted here that in the literature, Drucker-Prager criterion has been used by many authors to model the elasto-plastic behavior of intact rock matrix, see Pande & Xiong (1982) and Chen & Egger (1999) for example.

Although a modified Drucker-Prager yield function may be more suitable for this formulation in order to model plastic deformation properties of intact rock such as pressure dependency, strain hardening, transition from compressibility to dilatancy and stress path dependency (Chiarelli et al., 2003), the criterion given above may keep the formulation compact and does not lose generality. Other yield functions, such as the modified Drucker-Prager criterion (Chiarelli et al., 2003) or the modified Hoek-Brown criterion (Hoek et al., 1992), can also be integrated into the formulation without major mathematical difficulties.

With the researches conducted by Yuan & Harrison (2004) and Alejano & Alonso (2005), the decaying process of the rock dilatancy angle in line with plasticity can be described by the following negative exponential expression through the equivalent plastic strain of rock matrix, $\bar{\varepsilon}_R^p$ (Lai, 2002):

$$\psi_R = \psi_R^{peak} \exp(-r_R \bar{\varepsilon}_R^p)$$

(66)

where $r_R \geq 0$ is a parameter for modelling the decaying process of the dilatancy angle, and ψ_R^{peak} is the peak dilatancy angle of rock matrix and the following expression has been proposed by recovering the shape of the peak dilatancy angle of fractures given by Barton & Bandis (1982) and by assuming $\psi_R^{peak} = \varphi_R$ for null confinement pressures (Alejano & Alonso, 2005):

$$\psi_R^{peak} = \frac{\varphi_R}{1 + \log_{10}\sigma_c} \log_{10} \frac{\sigma_c}{-\sigma_3' + 0.1}$$

(67)

where σ_c is the unconfined compressive strength for intact rock. By Eqs. (66) and (67), the dependencies of rock dilatancy on plasticity, confining stress and scale are produced.

The equivalent plastic strain $\bar{\varepsilon}^p$ is computed by the following:

$$\bar{\varepsilon}^p = \int d\bar{\varepsilon}^p = \int \sqrt{\frac{2}{3}} d\varepsilon^p : d\varepsilon^p$$

(68)

Similarly, $C^{F,ep}$ in Eq. (58) is the elasto-plastic modulus tensor of fth set of fractures measured in the $X_1 X_2 X_3$ system, which can be calculated from its corresponding elastoplastic modulus tensor measured in the $x_1^f x_2^f x_3^f$ system, $C^{f,ep}$, with the assumption of small strain and by imposing the following tensor transformation:

$$C_{ijkl}^{F,ep} = l_{mi}^f l_{nj}^f l_{ok}^f l_{pl}^f C_{mnop}^{f,ep}$$

(69)

with

$$\mathbf{C}^{f,ep} = \mathbf{C}^f - \frac{\mathbf{C}^f : \dfrac{\partial Q_f}{\partial \boldsymbol{\sigma}'} \otimes \dfrac{\partial F_f}{\partial \boldsymbol{\sigma}'} : \mathbf{C}^f}{\dfrac{\partial F_f}{\partial \boldsymbol{\sigma}'} : \mathbf{C}^f : \dfrac{\partial Q_f}{\partial \boldsymbol{\sigma}'} + H_f}$$

(70)

where Cf is the fourth-order tangential elastic modulus tensor of the fth set of fractures, with $C^f_{3333} = s_f k_{nf}$, $C^f_{2323} = C^f_{3131} = s_f k_{sf}$, and with all other elements equal to zero. The symbols k_{nf}, k_{sf} and s_f are the normal stiffness, the tangential stiffness and the spacing of the fth set of fractures, respectively. The expressions for the elements in Cf mean that the strain of fractures is evaluated over the fracture spacing, not over the fracture aperture, thus enabling the proposed model to consider the post-sliding plasticity of fractures and nonlinear variations of k_{nf} and k_{sf} with dilatancy caused by shear loading, without violating the small strain assumption.

F_f, Q_f and H_f in Eq. (70) are the yield function, the plastic potential function and the hardening modulus of the fth set of fractures, respectively. The elasto-plastic behavior of the fractures is treated in a similar fashion as that for the rock matrix, with a non-associative Mohr-Coulomb criterion:

$$F_f = \sqrt{\tau^2_{zxf} + \tau^2_{zyf}} + \sigma'_{zf} \tan\varphi_f - c_f = 0$$

(71)

$$Q_f = \sqrt{\tau^2_{zxf} + \tau^2_{zyf}} + \sigma'_{zf} \tan\psi_f$$

(72)

where σ'_{zf}, τ_{zxf} and τ_{zyf} are the effective normal stress and the shear stresses on the fracture surfaces, respectively. c_f, φ_f and ψ_f are the cohesion, the friction angle and the mobilized dilatancy angle of the fth set of fractures, respectively. Similar to Eq. (66), ψ_f is also a shrinking function of the equivalent plastic strain of fractures $\bar{\varepsilon}^p_f$, and depends on normal stress and scale as well, in the following form:

$$\psi_f = \psi^{peak}_f \exp(-r_f \bar{\varepsilon}^p_f)$$

(73)

where r_f is the decaying parameter and ψ^{peak}_f is the peak dilatancy angle of the fth set of fractures, respectively, with the latter calculated by Eq. (26).

Thus at any loading step, as long as the stress increment of the equivalent rock mass, $d\sigma'$, is obtained, the local strain pertinent to fth set of fractures can be derived as follows:

$$d\varepsilon^F = (C^{F,ep})^{-1}:d\sigma' \tag{74}$$

and

$$d\varepsilon_{ij}^f = l_{im}^f l_{jn}^f d\varepsilon_{mn}^F \tag{75}$$

The separation of the incremental strain of fractures from that of the rock mass through the proposed equivalent constitutive model plays a significant role in the present study. It enables the formulation of strain-dependent hydraulic conductivity that accounts for the mobilized dilatancy behavior, which will be demonstrated in the following section.

Strain-Dependent Hydraulic Conductivity Tensor for Fractured Rocks

Consider a domain of flow that has been discretized into several sub-domains according to rock quality classification. Suppose that each sub-domain contains n sets of fractures, with average initial aperture b_{f0} and spacing s_f for the fth set of fractures. Starting from Eq. (22) and using the averaging concept for the hydraulic conductivity over the whole sub-domain, the equivalent initial hydraulic conductivity of the fth set of fractures, k_{f0}, in the examined sub-domain can be represented as (Castillo, 1972; Liu et al., 1999)

$$k_{f0} = \varsigma \frac{g b_{f0}^3}{v s_f} \tag{76}$$

where ς, as pointed out before, is a dimensionless constant introduced to penalize the real water conducting capacity of natural fractures with rough walls, finite scales, asperity areas and filling materials. The validity of using a constant value of ς has been examined by Zhou et al. (2006).

Assuming that the change in spacing s_f during modeling is negligible, under normal and shear stress loadings we have

$$k_f = \varsigma \frac{g b_f^3}{v s_f} = \varsigma \frac{g(b_{f0} + \Delta b_f)^3}{v s_f} \tag{77}$$

where Δb_f and k_f are the increment of the aperture and the equivalent hydraulic conductivity of the fth set of fractures under loading, respectively. Suppose that strain localization (Lai, 2002; Vajdova, 2003) is not dominantly exhibited in the concerned fractures, it is approximately valid that

$$\Delta b_f = s_f \Delta \varepsilon_{zf}$$

(78)

where $\Delta \varepsilon_{zf}$ is the increment of the normal strain of the fth set of fractures, which can be directly obtained from Eq. (75).

Substituting Eq. (78) into Eq. (77) then yields

$$k_f = k_{f0} \left(1 + \frac{s_f}{b_{f0}} \Delta \varepsilon_{zf} \right)^3$$

(79)

Following the theory proposed by Snow (1969), a strain-dependent equivalent hydraulic conductivity tensor for fractured rock masses with n sets of fractures is represented by

$$\mathbf{K} = \sum_f k_f (\boldsymbol{\delta} - \mathbf{n}_f \otimes \mathbf{n}_f) = \sum_f k_{f0} \left(1 + \frac{s_f}{b_{f0}} \Delta \varepsilon_{zf} \right)^3 (\boldsymbol{\delta} - \mathbf{n}_f \otimes \mathbf{n}_f)$$

(80)

where K is the equivalent hydraulic conductivity tensor of the examined rock mass, and n_f is the unit vector normal to the fth set of fractures.

The following significant implications can be observed from the formulation of K in Eq. (80):

- K is a cubic function of $\Delta \varepsilon_{zf}$, and any variation in ε_{zf} under loading will trigger the change in K, even in orders of magnitude. This exactly accounts for the coupling effect of mechanical loading (strain/stress) on hydraulic properties.

- K depends on incremental strains, rather than on stresses, which makes it possible to integrate various material nonlinearities in hydro-mechanical coupling analysis.

- In addition to cubic relation, the influence of $\Delta \varepsilon_{zf}$ on K is amplified by s_f / b_{f0}, indicating that K can be rather sensitive to b_{f0} and s_f. Therefore, techniques for estimating b_{f0} and s_f need to be carefully developed, on the basis of laboratory or in-situ hydraulic test data.

- The orientations of fractures possibly render K highly anisotropic, even if K is initially assumed isotropic, as has been systematically examined, e.g. by Liu et al. (1999).

- When implemented in a FEM code, a different K can be associated to each geological sub-domain or even to each element, as long as k_{f0}, b_{f0} and s_f for the sub-domains or elements can be estimated in advance.

- As a nature of the homogenized equivalent continuum approach, the size effect of fractures, especially the size-dependency of aperture, is not fully considered in the formulation of K for simplicity, even though it can be reflected to some degree through ς and scaled JRC and JCS values. The connectivity and the intersection effect of fractures, on the other hand, may have a more significant influence on K, but similarly, they cannot be properly considered in the equivalent continua without explicit representation of fractures. A rough remedy is to process the fracture system in such a way that only the connected fracture populations are included for conducting analyses.

To determine K of a fractured rock under any loading paths, a coupled hydro-mechanical process has to be invoked. With the assumption of incompressible rock matrix and fluid (e.g. groundwater), the governing equations for the coupled process of saturated fluid flow and deformation are given below as balance equation, geometric equation and fluid flow equation, respectively:

$$\sigma'_{ij,j} - \alpha p_{,i} + f_i = 0 \tag{81}$$

$$\varepsilon_{ij} = \frac{1}{2}\left(u_{i,j} + u_{j,i}\right) \tag{82}$$

$$\frac{\partial}{\partial x_i}\left(k_{ij}\frac{\partial h}{\partial x_j}\right) = \frac{\partial \varepsilon_v}{\partial t} \tag{83}$$

where f_i and u_i are the components of the body force and displacement in the ith direction, $h=p/\gamma_w+z$ the water head, z the vertical coordinate, γ_w the unit weight of water, and ε_v the volume strain of the rock mass.

In the coupled process given above, mechanical loading or disturbance to the rock mass results in change in flow properties and flow behavior through Eqs. (80) and (83), while the change in flow behavior leads to change in mechanical response of the rock mass through Eq. (81). When the coupled process reaches a stable state, the solution to K is also available.

Now we briefly discuss how to determine k_{f0}, b_{f0} and s_f in Eq. (80) based on laboratory or insitu hydraulic test or site investigation data. Obviously, the initial hydraulic conductivity, k_{f0}, can be determined by in-situ hydraulic tests. Suppose the initial hydraulic conductivity tensor, K_0, is known through in-situ hydraulic test, as suggested by Hsieh & Neuman (1985), then K_0 can be rewritten, from Eq. (80), in the following form:

$$\mathbf{K}_0 = \sum_f k_{f0}(\boldsymbol{\delta} - \mathbf{n}_f \otimes \mathbf{n}_f)$$

(84)

By optimizing Eq. (84), k_{f0} (f=1, ..., n) can be estimated if the number of the sets of critically oriented fractures, n, is less than or equal to 6 (i.e. the number of the independent components of K_0), regardless K_0 is assumed to be isotropic or anisotropic.

The average spacing of the fth set of fractures, s_f, can be roughly estimated from the statistics of drill holes or scanlines. An alterative, however, is to use RQD (Rock Quality Designation) for determining s_f, as suggested by Liu et al. (1999), when the value of RQD for a specific rock mass is known a priori.

After the initial hydraulic conductivity, k_{f0}, and the average spacing, s_f, of the fractures are determined, the mean initial aperture of the fractures, b_{f0}, is ready to be back-calculated from Eq. (76).

Validation of the Proposed Model

Hydraulic Conductivity of the Surrounding Rock of a Circular Tunnel in the Stripa Mine

Here we compare the proposed method with results from a previous study as presented by Liu el al. (1999) by applying the method to an excavated circular tunnel with a biaxial stress field, σ_x and σ_z. The physical model is illustrated in Fig. 18, which is actually a manifestation of the reality of the Stripa mine in Sweden (Kelsall et al., 1984; Pusch, 1989). The following description about the tunnel is directly taken from Liu et al. (1999):

A Buffer Mass Test was conducted in Stripa Mine over the period 1981-1985 (Kelsall et al., 1984; Pusch, 1989) to measure the permeability of a large volume of low permeability fractured rock mass by monitoring water flow into a 33 m long section of the tunnel, as a large scale in-situ experiment for the research and development programs of underground geological disposal of nuclear wastes of the participating countries of the Stripa Project. The radius of the tunnel is about 2.5 m with two major sets of fractures striking obliquely to the tunnel axis, as shown in Fig. 18.

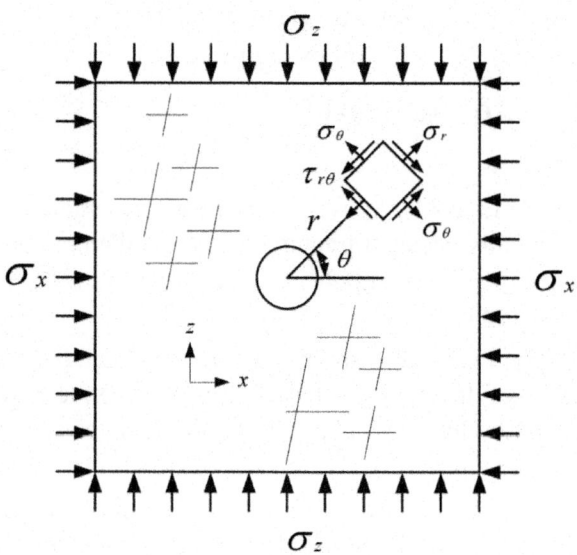

Figure 18. Sketch of a circular excavation in a biaxial stressed rock mass.

Fracture frequency measured in holes drilled from the tunnel was on average 4.5 fractures/m in inclined holes and 2.9 fractures/m in vertical holes. The initial stress field is anisotropic with high horizontal stress component and the conductivity of the virgin rock is about 10^{-10} m/s. The excavation of the test drift produced a dramatic increase in axial hydraulic conductivity in a narrow zone adjacent to the periphery of the drift. The conductivity increase is estimated to be 3 orders of magnitude.

The following assumptions are made in the calculations, with some of them similar to those in Liu et al. (1999):

- Statically uniform aperture and spacing distributions exist before excavation;

- Fracture spacing and continuity are not altered by the excavation;

- The high obliquity of the two major sets of fractures can be well approximated by two orthogonal sets of fractures;

- Excavation-induced strain redistribution may be adequately captured by the proposed equivalent elasto-plastic constitutive model.

Some of the parameters are directly taken from Liu et al. (1999), while other unavailable parameters are assumed, as listed in Table 5, in which the initial mechanical aperture of the fractures is back-calculated from Eq. (76) by taking $k_0 = 10^{-10}$ m/s. Consistent with Liu et al. (1999), the far-field stress components are taken as $\sigma_x = 20$ MPa and $\sigma_z = 10$ MPa, respectively.

Table 5. Geometrical and mechanical parameters for a circular tunnel

Category	Parameter	Setting
Intact rock matrix	Elastic modulus, E	37.5 GPa
	Poisson's ratio, v	0.25
	Cohesion, c_R	5 MPa
	Friction angle, φ_R	46°
Fractures	Initial mechanical aperture, b_0	0.0075 mm
	Spacing, s	0.27 m
	Normal stiffness, k_n	200 GPa/m
	Shear stiffness, k_s	100 GPa/m
	Dimensionless constant, ς	0.0067
	Cohesion, c_f	0.4 MPa
	Friction angle, φ_f	40°

To avoid the difficulty in determining the initial dilatancy angles and the corresponding decay parameters of fractures and intact rock matrix, associative flow rule is used in this simulation. Again for simplicity, both the normal stiffness and the shear stiffness of the fractures are assumed constant during excavation. The finite element mesh of the model is shown in Fig. 19, and the FEM program was run to simulate the excavation effect of the tunnel. Fig. 20 shows the deformation zone and plastic zone of the rock mass after the tunnel excavation. Fig. 21 plots the excavation-induced changes in hydraulic conductivities around the circular tunnel, which are directly compared with the results presented in Liu et al. (1999).

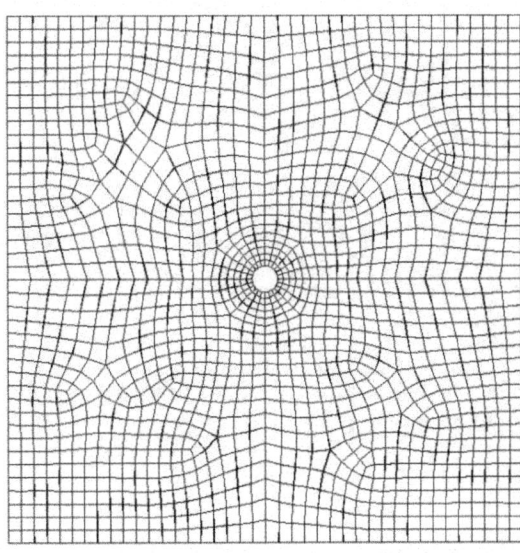

Figure 19. Finite element mesh for simulation of a tunnel excavation.

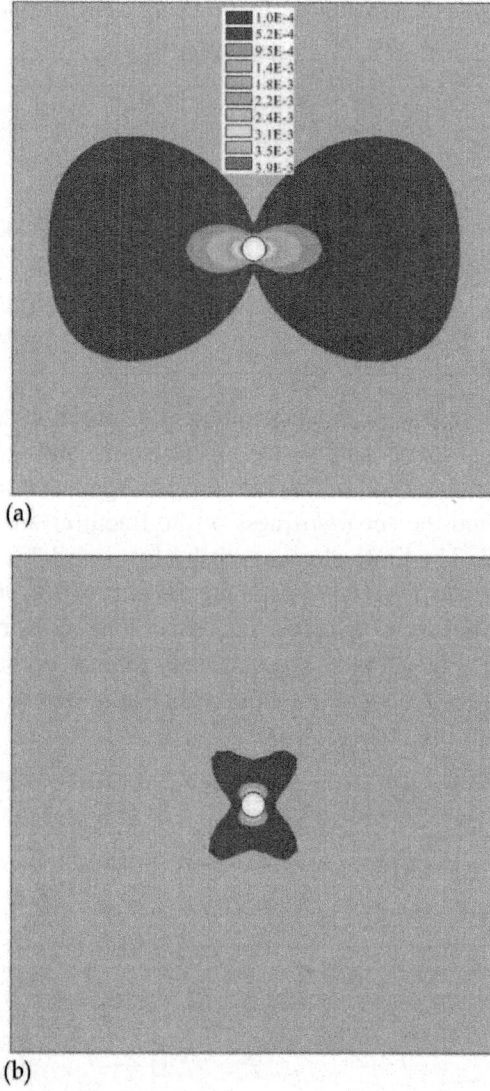

(a)

(b)

Figure 20. Deformation zone and plastic zone induced by the tunnel excavation: (a) deformation zone and (b) plastic zone.

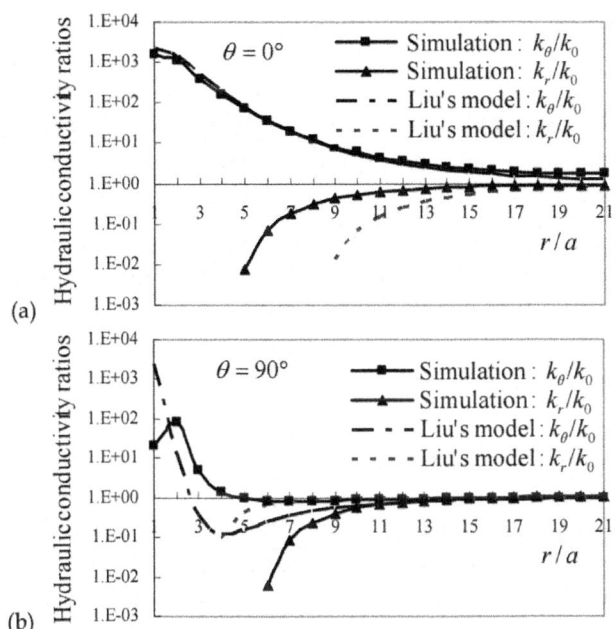

Figure 21. Excavation-induced hydraulic conductivity ratios around a circular tunnel in a biaxial stressed rock mass, where a is the radius of the tunnel and r is the distance away from the tunnel center. θ=0° denotes the horizontal direction while θ=90° the vertical direction.

It can be observed from Fig. 21 that generally tangential conductivities are found to increase greatly due to the formation of the excavation disturbed zone around the tunnel, while radial conductivities diminish greatly as a result of closure on related fractures. In the horizontal direction (i.e. θ=0°), the excavation-induced tangential hydraulic conductivity ratios, k_θ/k_0, predicted by our model are very close to the results presented in Liu et al. (1999). For radial hydraulic conductivity ratios, k_r/k_0, however, deviation occurs in the vicinity of the excavation. Such a deviation is also found both for k_θ/k_0 and for k_r/k_0 in the vertical direction (i.e. θ=90°).

Clearly, these deviations are largely resulted from the facts that (1) Different strain distribution patterns are assumed in the elastic model in Liu et al. (1999) and in our elastoplastic model; (2) Different methods are used to compute the strain increments of fractures. In Liu et al. (1999), normal strains of fractures were separated from rock matrix through a modulus reduction ratio empirically defined as a function of RMR, while in this simulation fracture strains were calculated by strain decomposition through an equivalent elasto-plastic constitutive model; (3) Radial and tangential fractures were assumed in Liu et al. (1999), leading to different background fracture networks; and (4) As mentioned above, some of the parameters, such as the shear strength of fractures and rock matrix, the shear stiffness and normal stiffness of the fractures, are unavailable in the literature (Kelsall et al., 1984; Pusch, 1989; Liu et al., 1999) and hence are empirically assumed in the calculations. If these parameters are determined based on in-situ or laboratory experiments, more convincing results may be achieved.

Despite the deviations, the trends of variation of the hydraulic conductivity ratios around the tunnel due to excavation are consistent between the two studies, and basically accord with the in-situ experimental observations, demonstrating the applicability of the present model in this section.

From Fig. 20, one observes that the excavation-induce deformation zone and plastic zone are asymmetric, due to the anisotropic initial stress field. As a result, the predicted hydraulic conductivities are highly anisotropic due to strain redistribution, as shown in Fig. 21. In the horizontal direction (i.e. $\theta=0°$), the deformation zone extends as far as more than 16 times of the tunnel radius and the plastic zone extends 2 times of the tunnel radius, while in the vertical direction (i.e. $\theta=90°$), they are, respectively, within 2 and 5 times of the tunnel radius. The asymmetry of deformation zone and plastic zone demonstrates why the predicted hydraulic conductivities approach k_0 more slowly in the horizontal direction than in the vertical direction. The changes in hydraulic conductivities resulted from strain redistribution in the disturbed rock mass indicate that a different hydraulic conductivity tensor should be associated to each geological sub-domain or even each element of the rock mass, which is important for hydro-mechanical coupling analyses.

Hydraulic Conductivity of a Cubic Block of Rock Mass with Three Orthogonal Sets of Identical Fractures

In this section, a numerical simulation is conducted to evaluate hydraulic behavior of a cubic block of rock mass containing three orthogonal sets of identical fractures under isotropic triaxial compression and shear loading. The primary goal is to investigate the change in the hydraulic conductivity of the rock mass with increasing shear load, which is obviously not achievable through any elastic models considering only the deformation of fractures under normal stresses, e.g. in Liu et al. (1999).

The underlying rock mass block model for examination, with a size of $10\times10\times10$ m (a scale that can represent both the initial mechanical and hydraulic REVs (Min et al., 2004)), is assumed to contain three orthogonal sets of identical fractures, as sketched in Fig. 22. The spacing, s, of each set of fractures and the initial aperture, b_0, of each fracture are assumed to be identical, with s=1 m and b_0=1 mm. The mechanical properties of each fracture are also regarded identical and for simplicity, both the normal stiffness and the shear stiffness of the fractures are assumed to be constant during shear loading. All parameters used in this simulation are listed in Table 6, and such parameter settings enable us to demonstrate how the hydraulic conductivity evolves from initial isotropy to anisotropy in the shearing process.

The examined rock mass block model is divided into 1000 brick elements, and the resultant mesh is shown in Fig. 22. The loading condition is as follows. First, triaxial compressive stresses are applied on the surfaces of the cubic block, with $\sigma_x=\sigma_y=\sigma_z$=20 MPa. Then, a shearing load, τ, is applied on the upper and lower surfaces of the block model step by step, increasing at an increment of 1 MPa until a maximum shear load, 20 MPa, is reached. At each step of shear loading, numerical divergence may occur. If numerical divergence does occur, the simulation program terminates after 1000 iterations with a modified NewtonRaphson method.

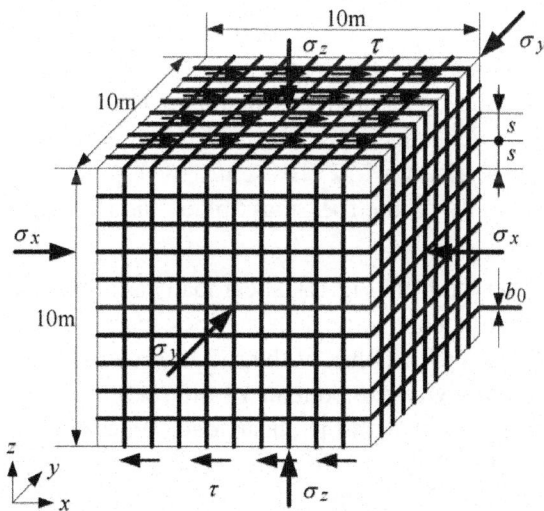

Figure 22. Sketch of a cubic block of rock mass with three orthogonal sets of identical fractures.

Table 6. Geometrical and mechanical parameters for a cubic block of fractured rock mass

Category	Parameter	Setting
Intact rock matrix	Elastic modulus, E	6 GPa
	Poisson's ratio, v	0.25
	Cohesion, c_R	1 MPa
	Friction angle, φ_R	46°
	Peak dilatancy angle, ψ_R^{peak}	35°
	Decay parameter of dilatancy, r_R	100
Fractures	Initial mechanical aperture, b_0	1 mm
	Spacing, s	1 m
	Normal stiffness, k_n	30 GPa/m
	Shear stiffness, k_s	10 GPa/m
	Dimensionless constant, ς	0.0067
	Cohesion, c_f	0.4 MPa
	Friction angle, φ_f	40°
	Peak dilatancy angle, ψ_f^{peak}	26°
	Decay parameter of dilatancy, r_f	100

Clearly, before the rock mass is loaded, its initial hydraulic properties are isotropic, with $k_{x0}=k_{y0}=k_{z0}=1.30\times10^{-2}$ cm/s by Eq. (84). Under the condition of isotropic compression, the rock mass remains elastic, the isotropic property of hydraulic conductivity is maintained, and the magnitude of the hydraulic conductivity reduces by 2 orders of magnitude due to compression of fractures, with $k_x=k_y=k_z=4.82\times10^{-4}$ cm/s by Eq. (80). When shear stress is added incrementally on the rock mass block model from 0 to 20 MPa, the proposed method

As can be observed from Fig. 23, shear load has a substantial impact on the evolution of hydraulic conductivity of the rock mass model. Before the shear load reaches 4 MPa, the response of the rock mass model remains elastic, and the hydraulic conductivity components of the rock mass model are basically identical and do not vary with the shear load. When the shear load exceeds 4 MPa, however, hydraulic conductivity of the model becomes anisotropic. Due to shear dilation of fractures in the z-direction, the major hydraulic conductivities parallel to the direction of shear load in x-y plane, k_x and k_y, increase mildly at first when the shear load is smaller than 8 MPa. Afterwards, they increase dramatically, reaching an increase of 3-4 orders of magnitude. They approach a relatively stable state after the shear load increases up to 14 MPa. Obviously, the increase of k_x and k_y is resulted from the dilatancy behavior of the fractures related to equivalent plastic strain, as shown in Fig. 24, where the mobilized dilatancy angle approaches zero as the shear load approaches 14 MPa. When the shear load exceeds 14 MPa, shear dilatancy of the related fractures becomes trivial and hence k_x and k_y become steady.

From Table 7 and Fig. 23, we can further see that k_x and k_y are very close to each other in values and they generally have the same varying trend with the increasing shear load.

Table 7. Major hydraulic conductivities of a cubic block of rock mass under isotropic compression and increasing shear loading

τ (MPa)	k_x (cm/s)	k_y (cm/s)	k_z (cm/s)	τ (MPa)	k_x (cm/s)	k_y (cm/s)	k_z (cm/s)
-	0.013016	0.013016	0.013016	10	0.279373	0.279350	0.000020
0	0.000482	0.000482	0.000482	11	1.088835	1.088816	0.000056
1	0.000482	0.000482	0.000482	12	2.204162	2.204158	0.000375
2	0.000482	0.000482	0.000482	13	3.171558	3.171559	0.001374
3	0.000483	0.000483	0.000482	14	3.676801	3.697449	0.022811
4	0.000494	0.000486	0.000474	15	3.915193	4.137786	0.224877
5	0.000543	0.000509	0.000444	16	4.063688	4.696511	0.635383
6	0.000657	0.000576	0.000372	17	4.243447	5.407600	1.167070
7	0.000742	0.000643	0.000282	18	4.635512	6.233203	1.600997
8	0.000704	0.000581	0.000207	19	5.390907	7.316177	1.928768
9	0.012562	0.012459	0.000106	20	6.462514	8.618240	2.159053

With the increase of shear load from 4 to 20 MPa, the change in the major hydraulic conductivity vertical to the direction of shear load, k_z, is even more interesting. Before the shear load reaches 10 MPa, k_z decreases significantly with increasing shear load and manifests a shear contraction-like behavior. When the shear load further increases, shear dilatancy occurs and k_z increases drastically, with changes in as high as 4-5 orders of magnitude. k_z reaches a relatively stable state after the shear load increases up to 17 MPa, which is actually a critical loading point that numerical instability may occur.

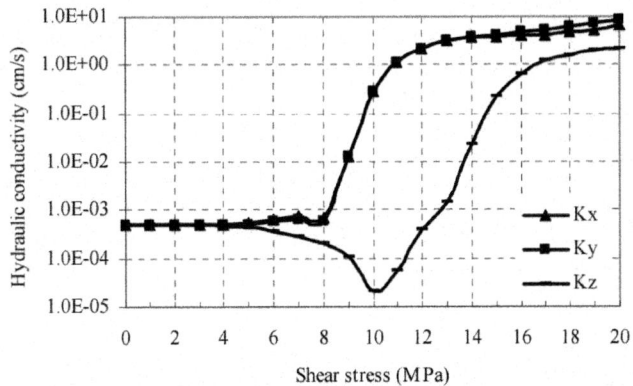

Figure 23. Major hydraulic conductivities of a cubic block of rock mass with increasing shear load.

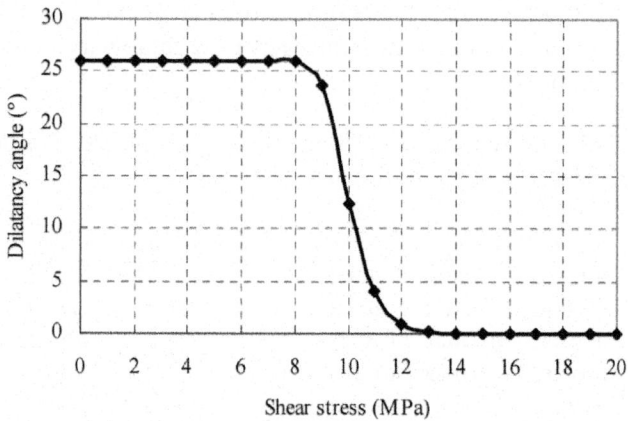

Figure 24. A typical case of mobilized dilatancy angle of a fracture with increasing shear load.

CONCLUSIONS

In this chapter, mathematical models were developed to estimate the hydraulic conductivity tensor for fractured rock masses subjected to mechanical loading or engineering disturbance. Emphases are placed on the investigation of the geological characteristics of rock masses as well as the coupling between fluid flow and stress/deformation, especially the effect of shear dilation or shear contraction on the hydraulic behavior of rock fractures.

The stress-dependent hydraulic conductivity tensor was formulated by using the superposition principle of flow dissipation energy on the basis of the concept of representative elementary volume (REV) and the assumption that rock masses can be treated as equivalent continuum media. The deformation behaviors of rock fractures subjected to normal and shear loadings are described with an elastic constitutive model, in which the pre-peak shear dilation or contraction of the fractures is empirically modelled. The validity of using the superposition principle of flow dissipation energy for development of the model is supported by the functional equivalence between the current formulation and the Snow's and Oda's models. This model is best suited for estimation of the hydraulic properties of rock masses at low stress level and with overall elastic response, and can be used to determine the applicability of the continuum approach to coupling analysis. The latter is achieved by performing numerical experiments to test the existence of the REV, and if exists, to further estimate the REV by gradually increasing the cubic volume of flow region, V_p, to see whether the hydraulic conductivity of the rock mass can eventually approach a steady point. The hydraulic properties and the REV size of the fractured rock mass at the construction site of the Laxiwa Hydropower Project were evaluated with the proposed model, and the calculation results were compared with the predictions of the Snow's model and validated by in-situ hydraulic tests, hence the feasibility of the proposed model in rock engineering practices is demonstrated.

The strain-dependent hydraulic conductivity tensor, on the other hand, was developed for disturbed rock masses under excavation or loading. In the model, a non-associative elasticperfectly plastic constitutive model was integrated to describe the deformation behaviors of the rock masses by characterizing them as equivalent continua containing one or multiple sets of parallel fractures. The clear advantages of the formulation are:

- The proposed hydraulic conductivity tensor is related to strains rather than stresses, hence enabling easier hydro-mechanical coupling analysis to include the effect of material nonlinearity of fractured rock masses.

- Beneficial from the equivalent non-associative elastic-perfectly plastic constitutive model, the hydraulic conductivity tensor considers the impact of shear dilatancy of fractures on fluid flow properties via mobilized dilatancy angles.

- When reduced to one dimensional case with a single fracture under normal and shear loadings, a closed-form solution to the hydraulic conductivity can be obtained, enabling validation of the model by laboratory coupled shear-flow tests of rock fractures.

- The proposed model is easy to be implemented in a FEM code, particularly suitable for numerical analysis of coupled hydro-mechanical processes in rock engineering.

The closed-form solution was validated by an existing coupled shear-flow test, and the evaluation results show that the proposed solution can closely describe the hydraulic behavior of a hard rock fracture under a wide range of normal and shear loads. The results of the simulation conducted to predict the excavation-induced hydraulic conductivities around a circular tunnel in a biaxial stress field at the Stripa mine are justified by in-situ experimental observations and compared with an existing elastic strain-dependent model, which show that engineering disturbance such as underground excavations may dramatically alter the hydraulic conductivities of the rock mass surrounding the excavations and change the isotropic pattern of the initial hydraulic conductivities. The numerical simulation on a cubic block model of a rock mass with three orthogonal sets of identical fractures under isotropic triaxial compression and shear loading further demonstrates that shear loading may drastically change the hydraulic properties of fractured rocks, in the magnitude of as high as 4-5 orders, and lead to high anisotropy of the hydraulic properties. Despite all these efforts, characterizing the hydraulic properties for fractured rock masses remains one of the most difficult research topics in rock mechanics. In the proposed models presented in this chapter, rock masses are assumed with rather regular distribution patterns of fractures, and the existence of a hydraulic conductivity tensor of the rock masses with any distribution of fractures is not discussed. The interaction between the fractures in the rock masses is also out of the scope of this chapter, and its effect on the hydraulic properties remains an open issue. Furthermore, the proposed models are established with a rather intuitive upscaling approach, and more rigorous homogenization schemes should be developed. All of there issues should be addressed in the future research.

ACKNOWLEDGEMENTS

The financial support from the National Natural Science Foundation of China (No. 51079107) and the National Natural Science Fund for Distinguished Young Scholars of China (No. 50725931), and the Program for New Century Excellent Talents in University (No. NCET-09-0610) for this study is gratefully acknowledged.

REFERENCES

1.   Alejano, L. R. & Alonso, E. (2005). Consideration of the dilatancy angle in rocks and rock masses. International Journal of Rock Mechanics and Mining Sciences, Vol. 42, No. 4, 481–507

2.   Barton, N. (1976). Rock mechanics review: the shear strength of rock and rock joints. International Journal of Rock Mechanics and Mining Sciences & Geomechanical abstracts, Vol. 13, No. 9, 255-279

3.   Barton, N. R. & Bandis, S. C. (1982). Effects of block size on the shear behavior of jointed rocks. In: Proc 23rd US Symp Rock Mechanics, Berkeley

4.   Barton, N.; Bandis, S. & Bakhtar, K. (1985). Strength, deformation and conductivity coupling of rock joints. International Journal of Rock Mechanics and Mining Sciences & Geomechanical abstracts, Vol. 22, No. 2, 121-140

5.   Bear, J. (1972). Dynamics of fluids in porous media. American Elsevier, New York

6.   Castillo, E. (1972). Mathematical model for two-dimensional percolation through fissured rock. In: Proc Int Symp Percolation through Fissured Rock, T1±D1–7, Stuttgart, Germany

7. Chen, Y. F.; Sheng, Y. Q. & Zhou, C. B. (2006). Strain-dependent permeability tensor for coupled M-H analysis of underground opening. Proceedings of the 4th Asian Rock Mechanics Symposium, pp. 271, Singapore, Nov 2006, World Scientific Publishing

8. Chen, Y. F.; Zhou, C. B. & Sheng, Y. Q. (2007). Formulation of strain-dependent hydraulic conductivity for fractured rock mass. International Journal of Rock Mechanics and Mining Sciences, Vol. 44, No. 7, 981-996

9. Chen, S. H. & Egger, P. (1999). Three dimensional elasto-viscoplastic finite element analysis of reinforced rock masses and its application. International Journal for Numerical and Analytical Methods in Geomechanics, Vol. 23, No. 1, 61–78

10. Chiarelli, A. S.; Shao, J. F. & Hoteit, N. (2003). Modeling of elastoplastic damage behavior of a claystone. Int J Plasticity, Vol. 19, 23–45

11. Esaki, T.; Du, S.; Mitani, Y.; Ikusada, K. & Jing, L. (1999). Development of a shear-flow test apparatus and determination of coupled properties for a single rock joint. International Journal of Rock Mechanics and Mining Sciences, Vol. 36, 641–50

12. Hoek, E.; Wood, D. & Shah, S. (1992). A modified Hoek-Brown criterion for jointed rock masses. In: Proc Rock Characterization Symp ISRM: Eurock 92, Hudson, J. A. (Ed.), 209–214, British Geotechnical Society, London

13. Hsieh, P. A. & Neuman, S. P. (1985). Field determination of the three-dimensional hydraulic conductivity tensor of anisotropic media. Water Resource Research, Vol. 21, No. 11, 1655–1665.

14. Huang, T. H.; Chang, C. S. & Chao, C. Y. (2002). Experimental and mathematical modeling for fracture of rock joint with regular asperities. Eng Fract Mech, Vol. 69, 1977– 1996

15. Indelman, P. & Dagan, G. (1993). Upscaling of permeability of anisotropic heterogeneous formations. Water Resources Research, Vol. 29, No. 4, 917-923

16. Jing, L. (2003). A review of techniques, advances and outstanding issues in numerical modeling for rock mechanics and rock engineering. International Journal of Rock Mechanics and Mining Sciences, Vol. 40, No., 283-353

17. Jing, L.; Stephansson, O. & Nordlund, E. (1993). Study of rock joints under cyclic loading conditions. Rock Mechanics and Rock Engineering, Vol. 26, No. 3, 215–32

18. Kelsall, P. C.; Case, J. B. & Chabannes, C. R. (1984). Evaluation of

excavation-induced changes in rock permeability. Int J Rock Mech Min Sci & Geomech Abstr, Vol. 21, No. 3, 123–35

19. Lai, T. Y. (2002). Multi-scale finite element modeling of strain localization in geomaterials with strong discontinuity. Ph.D. thesis, Stanford University

20. Liu, C. H.; Chen, C. X. & Fu, S. L. (2002). Testing study on seepage characteristics of single fracture with sand under shearing displacement. Chinese Journal of Rock Mechanics and Engineering, Vol. 21, No. 10, 1457-1461

21. Liu, J.; Elsworth, D. & Brady, B. H. (1999). Linking stress-dependent effective porosity and hydraulic conductivity fields to RMR. International Journal of Rock Mechanics and Mining Sciences, Vol. 36, 581-596

22. Liu, S. H. (1996). Generation of flow network and field tests on hydraulic conductivity for fractured rock mass. Northwestern Hydropower, Vol. 55, No. 1, 21-27

23. Lomize, G. M. (1951). Flow in fractured rocks. Gosenergoizdat, Moscow

24. Long, J. C. S.; Remer, J. S.; Wilson, C. R. & Witherspoon, P. A. (1982). Porous media equivalents for networks of discontinuous fractures. Water Resource Research, Vol. 18, No. 3, 645–58

25. Louis, C. (1971). A study of groundwater flow in jointed rock and its influence on the stability of rock masses. Rock Mechanics Research Report, No. 10, Imperial College of Science and Technology, London, Maini, YNT

26. Min, K. B. & Jing, L. (2003). Numerical determination of the equivalent elastic compliance tensor for fractured rock masses using the distinct element method. International Journal of Rock Mechanics and Mining Sciences, Vol. 40, No. 6, 795-816

27. Min, K. B.; Rutqvist, J.; Tsang, C. F. & Jing, L. (2004). Stress-dependent permeability of fractured rock masses: a numerical study. International Journal of Rock Mechanics and Mining Sciences, Vol. 41, No. 7, 1191-1210

28. Oda, M. (1985). Permeability tensor for discontinuous rock masses. Geotechnique, Vol. 35, No. 4, 483-195

29. Oda, M. (1986). An equivalent continuum model for coupled stress and fluid flow analysis in jointed rock masses. Water Resources Research, Vol. 22, No. 13, 1845-1856

30. Olsson, R. & Barton, N. (2001). An improved model for hydromechanical coupling during shearing of rock joints. International Journal of Rock

Mechanics and Mining Sciences, Vol. 38, No. 3, 317-329

31. Pande, G. N. & Xiong, W. (1982). An improved multilaminate model of jointed rock masses. In: Numerical Models in Geomechanics, Dungar, R.; Pande, G. N. & Studer, J. A. (Ed.), 218–226, Bulkema, Rotterdam

32. Patir, N. & Cheng, H. S. (1978). An average flow model for determining effects of threedimensional roughness on hydrodynamic lubrication. ASME Journal of Lubrication Technology, Vol. 100, 12-17

33. Plesha, M. E. (1987). Constitutive models for rock discontinuities with dilatancy and surface degradation. International Journal for Numerical and Analytical Methods in Geomechanics, Vol. 11, 345–62

34. Pusch, R. (1989). Alteration of the hydraulic conductivity of rock by tunnel excavation. Int J Rock Mech Min Sci & Geomech Abstr, Vol. 26, No. 1, 79–83

35. Snow, D. T. (1969). Anisotropic permeability of fractured media. Water Resources Research, Vol. 5, No. 6, 1273-1289

36. Vajdova, V. (2003). Failure mode, strain localization and permeability evolution in porous sedimentary rocks. Ph.D. thesis, Stony Brook University

37. Wang, M. & Kulatilake, P. H. S. W. (2002). Estimation of REV size and three dimensional hydraulic conductivity tensor for a fractured rock mass through a single well packer test and discrete fracture fluid flow modeling. International Journal of Rock Mechanics and Mining Sciences, Vol. 39, 887-904

38. Yuan, S. C. & Harrison, J. P. (2004). An empirical dilatancy index for the dilatant deformation of rock. International Journal of Rock Mechanics and Mining Sciences, Vol. 41, 679–86

39. Zimmerman, R. W.; Kumar, S. & Bodvarsson, G. S. (1991). Lubrication theory analysis of the permeability of rough-walled fractures. International Journal of Rock Mechanics and Mining Sciences, Vol. 28, No. 4, 325-331

40. Zhou, C. B.; Chen, Y. F. & Sheng, Y. Q. (2006). A generalized cubic law for rock joints considering post-peak mechanical effects. In: Proc GeoProc2006, 188–197, Nanjing, China

41. Zhou, C. B.; Sharma, R. S.; Chen Y. F. & Rong, G. (2008). Flow-Stress Coupled Permeability Tensor for Fractured Rock Masses. International Journal for Numerical and Analytical Methods in Geomechanics, Vol. 32, 1289-1309

42. Zhou, C. B. & Xiong, W. L. (1996). Permeability tensor for jointed rock masses in coupled seepage and stress fields. Chinese Journal of Rock

Mechanics and Engineering, Vol. 15, No. 4, 338-344

43. Zhou, C. B. & Xiong, W. L. (1997). Influence of geostatic stresses on permeability of jointed rock masses. Acta Seismologica Sinica, Vol. 10, No. 2, 193-204

44. Zhou, C. B.; Ye, Z. T. & Han, B. (1997). A study on configuration and hydraulic conductivity of rock joints. Advances in Water Science, Vol. 8, No. 3, 233-239

45. Zhou, C. B. & Yu, S. D. (1999). Representative elementary volume (REV): a fundamental problem for selecting the mechanical parameters of jointed rock mass. Chinese Journal of Engineering Geology, Vol. 7, No. 4, 332-336

Chapter 6

MECHANICAL BEHAVIOR OF 3D CRACK GROWTH IN TRANSPARENT ROCK-LIKE MATERIAL CONTAINING PREEXISTING FLAWS UNDER COMPRESSION

Hu-Dan Tang[1,2], Zhen-De Zhu[1], Ming-Li Zhu[3], and Heng-Xing Lin[4]

[1]Key Laboratory of Ministry of Education for Geomechanics and Embankment Engineering, Institute of Safety and Disaster Prevention Engineering, Hohai University, Nanjing, Jiangsu 210098, China

[2]School of Civil Engineering, Henan Polytechnic University, Jiaozuo, Henan 454000, China

[3]School of Energy Science and Engineering, Henan Polytechnic University, Jiaozuo, Henan 454000, China

[4]Water Conservancy Project Planning and Design Departments, Shanghai Investigation Design & Research Institute Co. Ltd., Shanghai 200434, China

ABSTRACT

Mechanical behavior of 3D crack propagation and coalescence is investigated in rock-like material under uniaxial compression. A new transparent rock-like material is developed and a series of uniaxial compressive tests on low temperature transparent resin materials with preexisting 3D flaws are performed in laboratory, with changing values of bridge angle β (inclination between the inner tips of the two preexisting flaws) of preexisting flaws in specimens. Furthermore, a theoretical peak strength prediction of 3D cracks coalescence is given. The results show that the coalescence modes of the specimens are varying according to different bridge angles. And the theoretical peak strength prediction agrees well with the experimental observation.

INTRODUCTION

Most of the elastic-brittle materials contain different patterns of flaws. In general, the mechanical behavior of brittle materials may be affected by the micromechanical behavior of the defects. The evolution of cracks depends on the properties of cracks such as size, location, orientation, and loading condition. The propagation of cracks plays a vital role in predicting the breakage process of rock specimens [1–12]. As a rule, the fracture surface is perpendicular to the maximum tensile stress direction. The experimental and theoretical research have shown that microcracks developed in different ways, such as tensile cracks, mixture cracks (tensile cracks, and shear cracks), and shear cracks, and became closed, frictional sliding, intergranular propagating, and kink propagating [13–15]. In the crack evolution process of brittle materials containing preexisting flaws, usually two types of crack are observed, which are wing cracks originating from the tips of preexisting flaws and secondary cracks. Wing cracks are usually caused by tension, while secondary cracks may develop due to shear [16]. Wing cracks initiation in rocks is favored with respect to secondary cracks because of lower toughness of the materials in tension than in shear [17–20]. It is mainly expected that crack initiation follows the direction parallel to the maximum compressive load [21]. Many experiments have been conducted to study the crack initiation, propagation path, and eventual coalescence of the preexisting flaws in specimens made of various substance, including natural rocks or rock-like materials under tensile and compressive loadings [4, 22–24].

From the practical point of view, nearly all rock engineering projects involve, to a certain extent, construction of structures in or on rock masses, which contain different types of flaws. As underground excavations progress into deeper and more complex geological environments, the eventual and ultimate limitation in all mining is depth [24]. Excavation-induced macroscale fractures, such as roof fall, side wall slab, and rock burst [25–29], occur extensively in the side walls of underground working face. Understanding of the failure modes around cavities in brittle rocks under compressive loading conditions becomes more and more important in searching solutions to the problem that engineering meets.

Fracture propagation leading to rock failure is a very important topic in rock mechanics research. A number of studies have been done on two-dimensional models plate with preexisting flaws. Crack initiation, propagation, and coalescence have been subjects of intensive investigation in rock mechanics, both theoretically and experimentally. The first theoretical study on the growth of preexisting two-dimensional flaws was put forward by Griffith [30, 31]. Griffith [30] further introduced the concept of critical energy release rate and the crack tip stress intensity factor (K). Relating to the field of rock mechanics, many experimental studies have been conducted to investigate the crack initiation, propagation, and interaction [1–12, 32, 33]. A number of studies have been done on two-dimensional (2D) model plates with throughgoing preexisting fractures, but as is known rock masses contain some finite size of flaws (three-dimensional (3D) flaws) existing inside or on the surface of rock materials. In terms of rock experiments, due to the nontransparency of rock material, it is difficult to trace the initiation, propagation, and interaction of fractures within the rock. That is to say, the crack growth analyses based on 2D model may not truly reflect the real failure properties. Then some studies have been done on 3D specimens [5, 34–46]. In reality, preexisting fractures are 3D in nature.

Recently, several experiments according 3D crack evolution have been investigated at the Rock Mechanics Laboratory at Hong Kong Polytechnic University. Samples that were prepared in the experiments included a variety of real rocks, PMMA, cement, gypsum, and resin samples. All samples contained a preexisting flaw [40–44]. According to these experiments, both wink cracks and petal cracks initiated from preexisting flaw tips of PMMA and marble samples, and shell-like cracks emerged from the flaw tips of the two materials referred to above sometimes. At the same time, antiwing cracks (opposite to the wing cracks) were induced from the tips of preexisting flaw at a certain distance in compressive stress zone in gabbros specimens [42, 43]. Liu et al. conducted a series of experimental tests to study 3D cracks propagation progress of a single surface flaw under the conditions of biaxial compression [44], and a 3D acoustic emission (AE) location system was used [42, 44].

However, most of previous studies were focused on the mechanisms and experiments of crack initiation, propagation, and interaction according to 2D cracks. Although some of significant results have been achieved, there were some deviations between the research results and the truth due to the nature of the material itself, the mechanisms of propagation and coalescence of 3D internal flaws are still not clear until now, and no existing theoretical explanation of 3D crack evolution was given.

Therefore, we attempt here to give a more refined study on the pattern of 3D crack initiation, propagation, and coalescence of transparent materials like rocks. On the basis of previous studies, the modeling material used in the paper is improved by being randomly embedded inside transparent resin material, certain aggregates of different sizes, and then heterogeneous transparent materials are obtained and successfully deal with the disadvantages of transparent materials which are isotropic. Experimental studies have shown newly developed transparent nonhomogeneous material properties close to real rocks, for the study of internal crack of rock which is no doubt highly beneficial. Due to fine brittle and transparency of the material, the internal crack growth can be clearly seen. Then the crack extension of the materials containing two preexisting flaws is investigated under uniaxial compression, with changing rock bridge angles, and rock bridge area is defined as shown in Figure 1; different modes of crack coalescence are observed in the 3D preexisting flaws specimens. Another main purpose of the paper is to predict the peak strength of transparent rock-like material containing preexisting flaws.

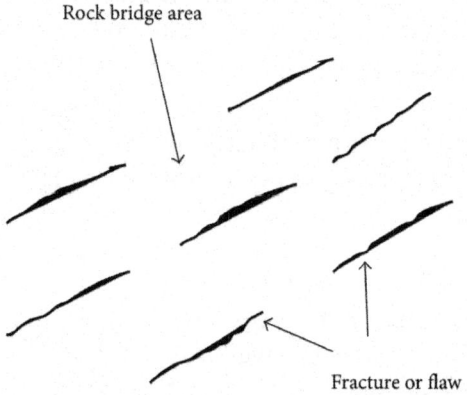

Figure 1: Rock bridge area in discontinuous rock.

SAMPLE PREPARATION AND EXPERIMENTAL TECHNIQUE

The discussion of the sample preparation and experimental technical contains three sections. The first section is the preparation of transparent casting resin modeling material; the second part is design of preexisting flaws in the samples; the third section is about the testing apparatus.

Preparation of Transparent Casting Resin Modeling Material Specimen

In the experiment, a new unsaturated resin is used to make specimens; sixty transparent rock-like parallelepiped samples are prepared and with cross section dimensions of 50 mm × 50 mm and a height of 100 mm are used. The mica sheet is fixed inside the mold through fine cotton according the needed angle. The precise calculation ratio of liquid resin is poured into the mold mica sheet fixed. Some aggregates with different particle sizes are randomly embedded inside the transparent resin material in the process of casting resin material modeling. At room temperature for 24 hours, the specimens are taken out from the mold. After repeatedly baking in the oven for 3 to 5 times, with each baking time about 30 minutes, the specimens are freezing to −30°C, and then this material is perfectly brittle, deforms without barreling, and has linear stress-strain behavior up to its burst-like fracture. The mechanical properties evaluated during the tests are as follows: Young's modulus E = 7.553 GPa; uniaxial compressive strength σ_c = 93.488 MPa; fracture toughness K_{IC} = 0.6 MPa·m$^{1/2}$.

Design of Preexisting Flaws

A thin mica film (the thickness of 0.1 mm) is used to model internal preexisting flaw during casting and be hold in the mold by cotton threads; it can represent a native open fracture of the rocks better for smaller stiffness than copper. The sizes of elliptical preexisting flaw are long axis $2c$ of 12 mm and short axis $2b$ of 8 mm. The positions and orientations of the slots are predetermined to give the inclination of the cracks (α = 30◦) and different rock bridge angle (β), which is the relative inclination between the cracks. For the sake of later discussions, the flaws are labeled as 1, 2. Three different bridge angles are used in the experiment, which are 60°, 85°, and 110°, as well as integrated species. Therefore, we can investigate the cracks coalescence along different rock bridge angles, as illustrated in Figure 2.

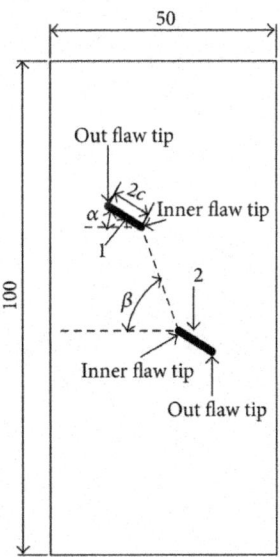

Figure 2: A specimen containing two preexisting flaws: the inclination is α, the rock bridge angle is β, the length of the preexisting flaw is $2c$, and the location of inner and outer flaw tips is defined.

Testing Apparatus

The uniaxial compression test is carried out with RMT-150B multifunction automatic rigid rock servo material testing machine (Figure 3). Displacement control mode is adopted as the load method in this experiment. The specimens are loaded to fail at a minimum loading speed of 0.01 mm/s. The loading system records the values of load, displacement, and other parameters and draws the curve of load-displacement instantaneously. A video camera is connected to the microscope and all the images are transferred to a computer instantaneously, so that the process of crack evolution can be analyzed conveniently after testing.

(a)

(b)

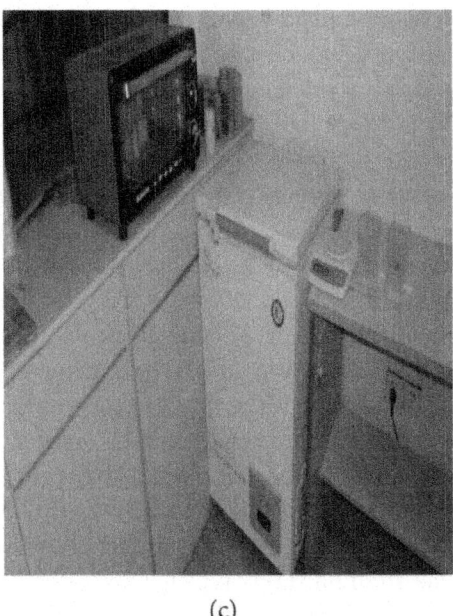

(c)

Figure 3: Test equipment: (a) data logger of RMT-150B multifunction automatic rigid rock servo material testing machine, (b) loading system of RMT-150B multifunction automatic rigid rock servo material testing machine, and (c) oven, which is used to test sample, make its curing as soon as possible, and increase its brittleness.

RESULTS AND ANALYSIS

Three types of models containing different rock bridge angles are tested to investigate the development of 3D fracture patterns. The following three sections depict the crack initiation, propagation, and coalescence of transparent resin materials with preexisting 3D cracks. The first section is general experimental observation; the second section is different model of crack coalescence for specimens containing different rock bridge angles; the third one is peak strength of 3D preexisting flaws specimens.

General Experimental Observation

Specimen with double preexisting flaws is experienced process of pressure elastic deformation, crack expansion, brittle failure, and residual strength on the whole. The coalescence of the specimen has much to do with the rock bridge angles.

Now the crack propagation process of specimen rock bridge angle 85° is described in detail. According to the loading record and images obtained in the loading process, first stage is pressure dense phase and then the elastic deformation; when the stress reaches about 50% of the peak strength, the crack initiation appears first as a sudden at the inner tips of preexisting crack 1 in the form of leaping and is about half the length of the prefabricated crack axis; the typical pattern of wing crack is shown in Figure 4(a). The wrapping wing cracks then start to curve around the preexisting flaw boundary. When the stress reaches about 60% of the peak strength, the crack emerges from the tips of preexisting flaw 2 as a sudden, and the length is roughly the same as the length of axis. With loading increasing, the wing crack emerges from lower tip of preexisting flaw 1 and the upper tip of preexisting flaw 2 and grows in a stable way; later, different from the results of 2D crack growing, antiwing wrapping crack (its growth direction is opposite to the wing wrapping crack) is induced from preexisting flaw 2, but the growing length is limited, as long as one-third of length of the short axis. At the same time, the wing cracks, respectively, from the upper tip of preexisting flaw 2 and lower tips of preexisting flaw 1 are growing towards each other but not coalescence. When the stress reaches about 70% of the peak strength, a tiny type tension crack turns up in the middle part of rock bridge area; ultimately the growing of the secondary crack and the propagation of wing cracks lead to the coalescence of crack induced by the preexisting flaws. When the stress reaches about 75% of the peak strength, cracks begin to grow from the upper tip of preexisting flaw 1. When the stress reaches about 90% of the peak strength, cracks come up in the no fissure zone and are quickly growing connecting with the cracks induced by the preexisting flaws. When the stress falls to about 20% of the peak strength, the effective bearing load area between particles is gradually reduced, and the specimen eventually damages, as shown in Figures 4(a) and 4(a′).

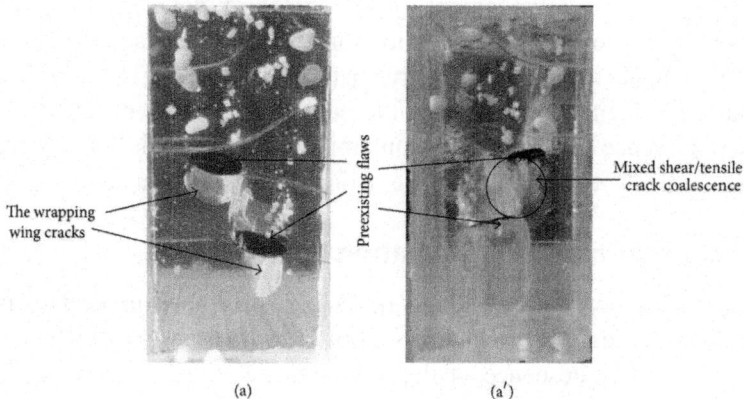

(a)   (a′)

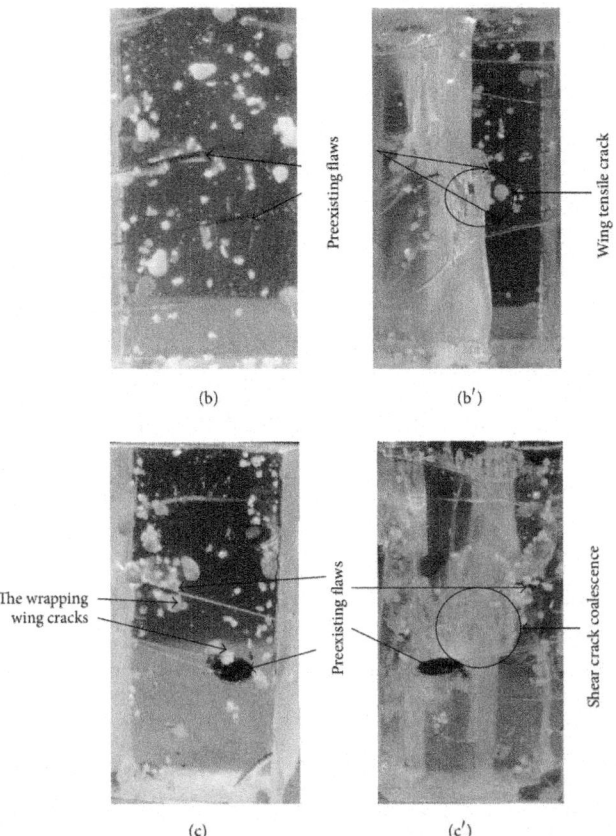

Figure 4: Modes of crack coalescence with different rock bridge angles: the inclination angle is about 30°; the frictional coefficient μ is about 0.577. (a) Showing the early stage of crack coalescence process under uniaxial compression when rock bridge angle is 85°, (a') showing the failure of the specimen under uniaxial compression when rock bridge angle is 85°; (b) showing the early stage of crack coalescence process under uniaxial compression when rock bridge angle is 110°, (b') showing the failure of the specimen under uniaxial compression when rock bridge angle is 110°; (c) showing the early stage of crack coalescence process under uniaxial compression when rock bridge angle is 60°, (c') showing the failure of the specimen under uniaxial compression when rock bridge angle is 60°.

The earlier stage of crack evolution of specimens with rock bridge angle 110° has little difference from the one with rock bridge angle 85°. The wrapping wing cracks all come up from the inner tips of the preexisting flaws. The difference is that no secondary cracks are produced in the area of Rock Bridge during the process of crack growing, but the eventual fracture is caused by wing cracks growing. That is to say, changing rock bridge angles will produce different mode of crack coalescence. As shown in Figures 4(b) and 4(b'), when loading is going on, the wing cracks start to curve towards the direction of loading, and wing crack plays a vital role in rock fracture.

The early stage of crack extension mode with bridge angle 60° resembles rock bridge angles 85° and 110° of the specimens; wing crack emerges from the inner and outer tips of crack 1 and crack 2 long axis one after another. When the stress reaches about 70% of the peak strength, secondary cracks emerge from the inner tips of crack 1 and crack 2, respectively. With loading increasing, when the stress reaches about 70% of the peak strength, cracks are growing quickly and begin coalescence in rock bridge area. Eventually they damage and form a shear failure surface, as shown in Figures 4(c) and 4(c').

In general, most cracks initiation appeared first at the inner tips of the preexisting flaws; then growth follows at the outer tips of the preexisting flaws, but some cracks initiation occurs in the reverse order, growth at the inner tips followed by cracks initiated at the outer tips. The growth of cracks at the outer tips is faster than that observed at inner tips. The types of cracking in rock bridge area can appear as either shear, tensile, or mix of both modes of crack coalescence. Shear cracks initiate in two different directions: coplanar or quasi-coplanar and oblique to the flaw [43]. A detailed discussion will be present in the next section.

THE MODES OF CRACK COALESCENCE IN ROCK BRIDGE AREA

In 2D modes, Wong and Chau [46] concluded that there were three modes of coalescence in rock bridge area. Patterns of crack coalescence of sandstone-like material containing two parallel inclined frictional cracks under uniaxial compression load are shown in Figure 5. The influence roles of the possible orientations of cracks included the values of inclination of preexisting cracks α, bridge angle β, and the frictional coefficient μ on the surfaces of the two preexisting cracks. When crack coalescence occurs, three main types of cracking can be identified in the rock bridge area: wing cracks, which are tensile in nature; secondary cracks, which are mainly shear in nature and are normally parallel to the preexisting cracks; mixed shear/tensile crack coalescence. In all, three main modes are as follows: S-mode (shear crack coalescence), M-mode (mixed shear/tensile crack coalescence), and W-mode (wing tensile crack coalescence), as can be seen in Figure 5. According to the loading record, our interest is placed on the coalescence pattern in the rock bridge area. When the bridge angle is 85° and when the stress reaches about 70% of the peak strength, a tiny secondary crack turns up in the middle part of rock bridge area; ultimately the growing of the secondary crack and the propagation of wing cracks lead to the coalescence of crack induced by the preexisting flaws. As can be seen in Figures 4(a) and 6(a), contrasting with the modes of 2D crack coalescence concluded by Wong, when $\alpha = 30°$, $\beta = 85°$, the observations are resemblance as the situation shown in Figure 5(b). And the mode is M-mode (mixed shear and tensile crack coalescence). When the rock bridge angle is 110°, under uniaxial compression load, wing cracks initiate and grow from tips of preexisting cracks. Wing cracks from the inner tip of crack 1 propagate downward to the outer tip of crack 2; at the same time, wing cracks from outer tip of crack 2 propagate upward to the inner tip of crack 1.

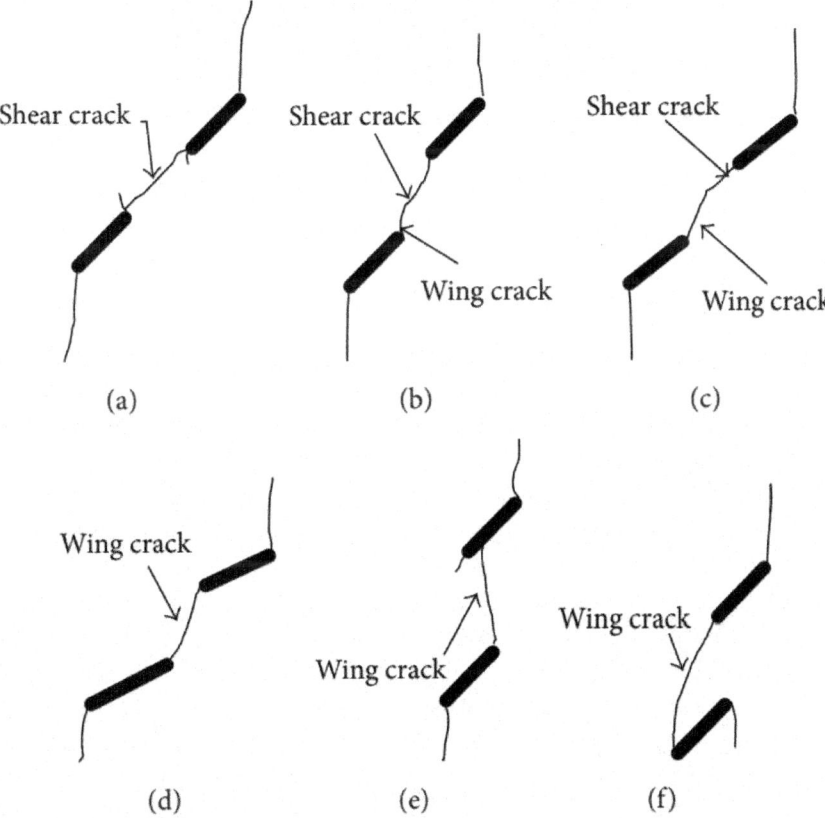

Figure 5: Six different patterns of crack coalescence were observed in the 2-flaw specimens. The notion of S, M, and W indicated the shear mode crack coalescence, mixed (shear/tensile) mode crack coalescence, and wing tensile mode crack coalescence (after Wong and Chau [46]).

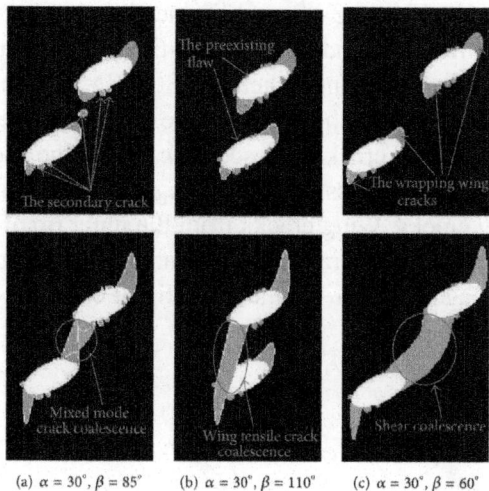

(a) $\alpha = 30°, \beta = 85°$ (b) $\alpha = 30°, \beta = 110°$ (c) $\alpha = 30°, \beta = 60°$

Figure 6: Three different patterns of 3D crack coalescence are observed in the tests.

However, the specimens failed by axial splitting rather than localized coalescence failure. As shown in Figures4(b) and 6(b), comparison with the coalescence mode of 2D crack induced which is proposed by Wong, seen in Figure 5(f), this crack coalescence mode is W-mode (wing tensile crack coalescence). When the rock bridge angle is about 60°, wing cracks nucleation at both inner and outer tips of the preexisting crack normally occurs first, but before the wing cracks propagate further, secondary shear cracks nucleate from both kinks at inner tips. The secondary cracks nucleate from both kinks at tips. The propagation of these secondary cracks leads to shear coalescence in the rock bridge area while wing cracks spread to the edges of the specimen, as shown in Figures 4(c) and 6(c), and shear failure surface is formed eventually. This kind of coalescence is mainly induced by a high shear stress concentration in the bridge area. Our observations suggest that whenever the two preexisting main cracks are in alignment, the shear interactions between the preexisting cracks become dominant. In contrast with the coalescence mode of 2D crack induced which is proposed by Wong, seen in Figures 4(c) and 6(c), this crack coalescence mode is S-mode (shear crack coalescence), but there are some differences that the dominant induced role is shear stress, but the effect of tensile cannot be neglected.

PEAK STRENGTH OF SPECIMEN WITH FLAWS

Peak strength prediction of rock containing preexisting flaws is discussed in this section. The mode raised by Ashby and Hallam [47] is employed. Ashby and Hallam derived the following approximate for wing cracks growing, which nucleated from a preexisting inclined crack of length $2c$ when the specimen was subject to uniaxial compression strength σ_1:

$$
\frac{K_I}{\sigma_1 \sqrt{\pi c}}
$$

$$
= \frac{(\sin 2\psi - \mu + \mu \cos 2\psi)}{(1 + L)^{3/2}} \left[0.23L + \frac{1}{\sqrt{3}(1 + L)^{1/2}} \right]
$$

$$
+ \left[\frac{2\varepsilon_0 (L + \cos \psi)}{\pi} \right]^{1/2},
$$

$$(1)$$

where σ_1 is the uniaxial compression strength, ψ is the angle measured from the σ_1-direction along the main surface of the flaw ($\psi = 90° - \alpha$), $2c$ is the length of the preexisting flaw, and the flaw density ε_0 is defined as Nc^2/A (N is the number of flaw per area A). Although strictly speaking (1) is for the case of multiple initial flaws, it was found that it also can be employed for the specimen containing two flaws. Thus, the peak uniaxial compressive strength σ_1^{max} of a flawed specimen can be estimated by Wong and Chau [46]:

$$
\sigma_1^{max} = \frac{K_{IC}}{\sqrt{\pi c}} \left\{ \frac{[\sin 2\psi - \mu + \mu \cos 2\psi]}{(1 + L_{cr})^{3/2}} \left[0.23L_{cr} \right. \right.
$$

$$
\left. + \frac{1}{\sqrt{3}(1 + L_{cr})^{1/2}} \right] + \left[\frac{2e_0 (L_{cr} + \cos \psi)}{\pi} \right]^{1/2} \right\}^{-1},
$$

$$(2)$$

where K_{IC} is the fracture toughness (in this paper $K_{IC} = 0.6$ MPa·m$^{1/2}$ for our modeling material), $L_{cr} = l_{max}/c$ ($l_{max} = 2b \sin \beta$ is the maximum possible value for length of the coalesced wing cracks, and $2b$ is the distance between the two flaws), and μ is the frictional coefficient along the main shear crack; the orientation of the shear crack for which the nucleation of the wing crack is most favorable is given by $2\psi = \tan^{-1}(1/\mu)$.

In this paper, the initial flaw density of specimen containing two flaws is $e_0 = 0.015$ ($\varepsilon_0 = Nc^2/A$; note that $N=2$, $A = 0.05$ m $\times$ 0.10 m, and $c = 0.004$ m). Predictions of the normalized peak strength $(\sigma_1^{max} \sqrt{\pi c}/K_{IC})$ by using (2) are listed in Table 1; furthermore the relationship between stress and strain of experimental results with different rock bridge angles is compared (see in Figure 7). As given in (1), the former part of the formula,

$$\frac{K_I}{\sigma_1 \sqrt{\pi c}}$$

$$= \frac{(\sin 2\psi - \mu + \mu \cos 2\psi)}{(1 + L)^{3/2}} \left[0.23L + \frac{1}{\sqrt{3} (1 + L)^{1/2}} \right]. \tag{3}$$

Equation (3) was derived by Ashby and Hallam, which is an approximate expression for mode 1 stress intensity factor K_I at the tip of the wing cracks, and the wing cracks nucleate from a preexisting inclined crack of length $2c$ when the solid is subject to uniaxial compression strength σ_1.

Table 1: Mechanical parameters of specimens with preexisting cracks of different bridge angles.

β (°)	Peak strength		$\varepsilon/10^{-3}$	E/GPa
	Experimental	Theoretical		
Complete specimen	21.46	—	16.08	6.15
60	13.13	10.09	15.12	5.33
85	12.86	16.51	14.04	5.15
110	12.71	13.68	14.65	5.21

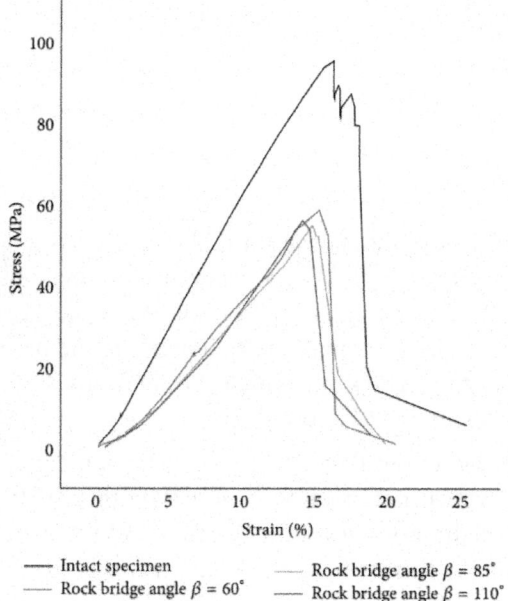

Figure 7: Stress and strain curves of samples with preexisting cracks of different bridge angles.

If peak strength is to be predicted, crack interaction and coalescence must be incorporated into the analysis. Using beam theory, the following K_I is due to crack interactions using beam theory, as can be seen from the later part of (2), and written as follows:

$$\frac{K_I}{\sigma_1 \sqrt{\pi c}} = \left\{ \frac{2e_0 (L + \cos \psi)}{\pi} \right\}^{1/2} .$$

(4)

Combining (3) and (4) gives the total stress intensity factor K_I for the wing cracks with crack interaction. Equation (3) completes the elastic theory for cracks. But, as known, rock materials can become plastic if the compressive stress is large enough. When a beam of thickness t and depth b is subjected to an axial stress σ_1 and a bending moment m, it starts to yield when the maximum surface stress reaches the yield strength. Hence, an additional contribution to stress intensity can be written as (4). But as shown in Figure 7, the transparent resin material undergoes elastic deformation dominantly; it suffers axial compression load but no significant bending. In other words, stress-strain curves are typical of brittle behavior: the nonlinear strain before peak strength is fairly small, and resistance drops dramatically afterwards. So the influence of beam is negligible, and the equation which can be applied in the study is (3). However, some modification has been made about the equation; that is, when $\alpha < 45°$, $|\cos 2\psi|$ should be applied. The experimental observations and theoretical results of peak strength of specimens are shown in Table 1.

The prediction by using the Ashby-Hallam model [46], which is description in the previous section, is presented here for comparison, as shown in Table 1; it is clear to see that the predicted theoretical peak strength agrees well with the experimental observation, but some deviations still exist in the modified model. For example, the intensity tendency does not perfectly agree with the result of the experiment. Furthermore, the Ashby-Hallam model should not be applied without modification when the inclination of preexisting flaws $\alpha < 45°$ and the modification to be made requires more detailed analysis in the future.

CONCLUSION

In this paper, experimental results on the mechanism of 3D crack propagation and coalescence as well as the peak strength of transparent rock-like material containing preexisting flaws under uniaxial compression are presented. The specimens used in this study are made of frozen transparent resin material with different rock bridge angles; the following is found:

- It can be observed that coalescence in 3D flaws with different rock bridge angles can be identified as the shear mode, the mixed mode (tensile mode and shear mode), and wing tensile mode. When the inclination angle $\alpha = 30°$ and frictional coefficient $\mu = 0.57$, the coalescence mode is dominated by different rock bridge angles. When $\beta = 60°$, shear mode coalescence occurs; when $\beta = 85°$, mixed mode coalescence occurs; when $\beta = 110°$, wing tensile mode coalescence occurs. Nevertheless, more 3D experimental and theoretical studies need to be carried out.

- The existence of flaws greatly reduces the compression strength of the specimen, and the cracks existing make the peak strengths reduced. The uniaxial peak strength prediction of 3D cracks by Ashby-Hallam [46] compares well with the experimental result. And there is some modification of the mode which has been made. Nevertheless, further modification should be done to give a better prediction of peak strength.

ACKNOWLEDGMENTS

The authors are grateful for the support of this work by the Natural Science Foundation of China (nos. 51404095, 51379065, and 41272329), the Chinese National Key Fundamental Research 973 Programme (2011CB013504), Colleges and Universities in Henan Province, the Construction of Deep Mine Open and Key Laboratory Open Fund (2013KF-06), the Education Department of Henan Province Science and Technology Research Projects (13B560040), and Scientific Research Foundation of Henan Polytechnic University, Dr. (B2011-105).

REFERENCES

1. Y. Ichikawa, K. Kawamura, K. Uesugi, Y.-S. Seo, and N. Fujii, "Micro- and macrobehavior of granitic rock: observations and viscoelastic homogenization analysis," Computer Methods in Applied Mechanics and Engineering, vol. 191, no. 1-2, pp. 47–72, 2001.

2. H. Haeri, K. Shahriar, M. F. Marji, and P. Moarefvand, "Cracks coalescence mechanism and cracks propagation paths in rock-like specimens containing pre-existing random cracks under compression,"Journal of Central South University, vol. 21, no. 6, pp. 2404–2414, 2014.

3. A. Bobet, "The initiation of secondary cracks in compression," Engineering Fracture Mechanics, vol. 66, no. 2, pp. 187–219, 2000.

4. R. H. C. Wong, K. T. Chau, C. A. Tang, and P. Lin, "Analysis of crack coalescence in rock-like materials containing three flaws—part I: experimental approach," International Journal of Rock Mechanics and Mining Sciences, vol. 38, no. 7, pp. 909–924, 2001.

5. E. Sahouryeh, A. V. Dyskin, and L. N. Germanovich, "Crack growth

under biaxial compression,"Engineering Fracture Mechanics, vol. 69, no. 18, pp. 2187–2198, 2002.

6. Y.-P. Li, L.-Z. Chen, and Y.-H. Wang, "Experimental research on pre-cracked marble under compression," International Journal of Solids and Structures, vol. 42, no. 9-10, pp. 2505–2516, 2005.

7. L. N. Y. Wong and H. H. Einstein, "Crack coalescence in molded gypsum and Carrara marble: part 1. Macroscopic observations and interpretation," Rock Mechanics and Rock Engineering, vol. 42, no. 3, pp. 475–511, 2009.

8. L. N. Y. Wong and H. H. Einstein, "Crack coalescence in molded gypsum and carrara marble: part 2—Microscopic observations and interpretation," Rock Mechanics and Rock Engineering, vol. 42, no. 3, pp. 513–545, 2009.

9. C. H. Park and A. Bobet, "Crack coalescence in specimens with open and closed flaws: a comparison,"International Journal of Rock Mechanics and Mining Sciences, vol. 46, no. 5, pp. 819–829, 2009.

10. T. Y. Ko, H. H. Einstein, and J. Kemeny, "Crack coalescence in brittle material under cyclic loading," inProceedings of the 41st US Symposium on Rock Mechanics, ARMA-06-930, Golden, Colo, USA, June 2006.

11. C. H. Park and A. Bobet, "Crack initiation, propagation and coalescence from frictional flaws in uniaxial compression," Engineering Fracture Mechanics, vol. 77, no. 14, pp. 2727–2748, 2010.

12. C.-a. Tang and Y.-f. Yang, "Crack branching mechanism of rock-like quasi-brittle materials under dynamic stress," Journal of Central South University, vol. 19, no. 11, pp. 3273–3284, 2012.

13. H. Li and L. N. Y. Wong, "Influence of flaw inclination angle and loading condition on crack initiation and propagation," International Journal of Solids and Structures, vol. 49, no. 18, pp. 2482–2499, 2012.

14. H. Haeri, K. Shahriar, M. F. Marji, and P. Moarefvand, "A coupled numerical-experimental study of the breakage process of brittle substances," Arabian Journal of Geosciences, vol. 8, no. 2, pp. 809–825, 2015.

15. C. Nielsen and S. Nemat-Nasser, "Crack healing in cross-ply composites observed by dynamic mechanical analysis," Journal of the Mechanics and Physics of Solids, vol. 76, pp. 193–207, 2015.

16. K. Horii, R. Yamada, and S. Harada, "Strength deterioration of nonfractal particle aggregates in simple shear flow," Langmuir, vol. 31, no. 29, pp. 7909–7918, 2015.

17. R. H. C. Wong, C. A. Tang, K. T. Chau, and P. Lin, "Splitting failure in brittle rocks containing pre-existing flaws under uniaxial compression," Engineering Fracture Mechanics, vol. 69, no. 17, pp. 1853–1871, 2002.

18. B. Shen, O. Stephansson, H. H. Einstein, and B. Ghahreman, "Coalescence of fractures under shear stresses in experiments," Journal of Geophysical Research, vol. 100, no. 4, pp. 5975–5990, 1995.

19. H. Jiefan, C. Ganglin, Z. Yonghong, and W. Ren, "An experimental study of the strain field development prior to failure of a marble plate under compression," Tectonophysics, vol. 175, no. 1–3, pp. 269–284, 1990.

20. J. T. Miller and H. H. Einstein, "Crack coalescence tests on granite," in Proceedings of the 42nd US Rock Mechanics Symposium (USRMS '08), ARMA-08-162, San Francisco, Calif, USA, June 2008.

21. L. N. Y. Wong and H. H. Einstein, "Using high speed video imaging in the study of cracking processes in rock," Geotechnical Testing Journal, vol. 32, no. 2, pp. 164–180, 2009.

22. S. Nemat-Nasser and H. Horii, "Compression-induced nonplanar crack extension with application to splitting, exfoliation, and rockburst," Journal of Geophysical Research, vol. 87, no. 8, pp. 6805–6821, 1982.

23. S. Q. Yang, Y. H. Dai, L. J. Han, and Z. Q. Jin, "Experimental study on mechanical behavior of brittle marble samples containing different flaws under uniaxial compression," Engineering Fracture Mechanics, vol. 76, no. 12, pp. 1833–1845, 2009.

24. R. J. Fowell and C. Xu, "The use of the cracked Brazilian disc geometry for rock fracture investigations,"International Journal of Rock Mechanics and Mining Sciences and, vol. 31, no. 6, pp. 571–579, 1994.

25. S. L. Crouch, "Analysis of stresses and displacements around underground excavations: an application of the displacement discontinuity method," University of Minnesota Geomechanics Report, University of Minnesota, Minneapolis, Minn, USA, 1967.

26. E. Hoek and E. T. Brown, Underground Excavations in Rock, Institute of Mining and Metallurgy, London, UK, 1980.

27. B. G. White, "Shear mechanism for mining-induced fractures applied to rock mechanics of coal mines," in Proceedings of the 21st International Conference on Ground Control in Mining, pp. 328–334, West Virginia University, Morgantown, WVa, USA, 1999.

28. R. T. Ewy and N. G. W. Cook, "Deformation and fracture around cylindrical openings in rock—I. Observations and analysis of deformations," International Journal of Rock Mechanics and Mining

Sciences & Geomechanics Abstracts, vol. 27, no. 5, pp. 387–407, 1990.

29. R. T. Ewy and N. G. W. Cook, "Deformation and fracture around cylindrical openings in rock-II. Initiation, growth and interaction of fractures," International Journal of Rock Mechanics and Mining Sciences and, vol. 27, no. 5, pp. 409–427, 1990.

30. A. A. Griffith, "The phenomena of rupture and flow in solids," Philosophical Transactions of the Royal Society of London Series A, vol. 221, pp. 163–198, 1921.

31. A. A. Griffith, "The theory of rupture," in Proceedings of the 1st International Congress for Applied Mechanics, pp. 55–63, Delft, The Netherlands, April 1924.

32. R. L. Kranz, "Crack-crack and crack-pore interactions in stressed granite," International Journal of Rock Mechanics and Mining Sciences & Geomechanics Abstracts, vol. 16, no. 1, pp. 37–47, 1979.

33. M. L. Batzle, G. Simmons, and R. W. Siegfried, "Microcrack closure in rocks under stress: direct observation," Journal of Geophysical Research, vol. 85, no. 12, pp. 7072–7090, 1980.

34. A. V. Dyskin, R. J. Jewell, H. Joer, E. Sahouryeh, and K. B. Ustinov, "Experiments on 3-D crack growth in uniaxial compression," International Journal of Fracture, vol. 65, no. 4, pp. R77–R83, 1994.

35. A. V. Dyskin, E. Sahouryeh, R. J. Jewell, H. Joer, and K. B. Ustinov, "Influence of shape and locations of initial 3-D cracks on their growth in uniaxial compression," Engineering Fracture Mechanics, vol. 70, no. 15, pp. 2115–2136, 2003.

36. A. V. Dyskin, L. N. Germanovich, R. J. Jewell, H. Joer, J. S. Krasinski, and K. K. Lee, "Study of 3-D mechanisms of crack growth and interaction in uniaxial compression," ISRM News Journal, vol. 2, no. 1, pp. 17–20, 1994.

37. A. Srivastava and S. Nemat-Nasser, "Overall dynamic properties of three-dimensional periodic elastic composites," The Royal Society of London—Series A: Proceedings, vol. 468, no. 2137, pp. 269–287, 2012.

38. C. K. Teng, X. C. Yin, and S. Y. Li, "An experimental investigation on 3D fractures of non-penetrating crack in plane samples," Acta Oceanologica Sinica, vol. 30, no. 4, pp. 371–378, 1987 (Chinese).

39. X. C. Yin, S. Y. Li, and H. Li, "Experimental study of interaction between two flanks of closed crack," Acta Geophysica Sinica, vol. 31, no. 3, pp. 307–314, 1988 (Chinese).

40. R. H. C. Wong, M. L. Huang, M. R. Jiao, C. A. Tang, and W. Zhu, "The

mechanisms of crack propagation from surface 3-D fracture under uniaxial compression," Key Engineering Materials, vol. 261, no. I, pp. 219–224, 2004.

41. R. H. C. Wong, Y. S. Guo, and L. Y. Li, "Anti-wing crack growth from surface flaw in real rock under uniaxial compression," in Fracture of Nano and Engineering Materials and Structures: Proceedings of the 16th European Conference of Fracture, Alexandroupolis, Greece, July 3–7, 2006, E. E. Gdoutos, Ed., pp. 825–826, Springer, Amsterdam, The Netherlands, 2006.

42. R. H. C. Wong, Y. S. Guo, and K. T. Chau, "The fracture mechanism of 3D surface fault with strain and acoustic emission measurement under axial compression," Key Engineering Materials, vol. 358, pp. 2360–3587, 2007.

43. Y. S. Guo, R. H. C. Wong, W. S. Zhu, K. T. Chau, and S. Li, "Study on fracture pattern of open surface-flaw in gabbro," Chinese Journal of Rock Mechanics and Engineering, vol. 26, no. 3, pp. 525–531, 2007.

44. L. Q. Liu, P. X. Liu, H. C. Wong, S. P. Ma, and Y. S. Guo, "Experimental investigation of three-dimensional propagation process from surface fault," Science in China, Series D: Earth Sciences, vol. 51, no. 10, pp. 1426–1435, 2008.

45. Y. S. Guo, The study on experiment, theory and numerical simulation of fracture of three-dimensional flaws in brittle materials [Ph.D. thesis], Shandong University, Jinan, China, 2007.

46. R. H. C. Wong and K. T. Chau, "Crack coalescence in a rock-like material containing two cracks,"International Journal of Rock Mechanics and Mining Sciences, vol. 35, no. 2, pp. 147–164, 1998.

47. M. F. Ashby and S. D. Hallam, "The failure of brittle solids containing small cracks under compressive stress states," Acta Metallurgica, vol. 34, no. 3, pp. 497–510, 1986.

Chapter 7

APPLICATION OF BASE FORCE ELEMENT METHOD ON COMPLEMENTARY ENERGY PRINCIPLE TO ROCK MECHANICS PROBLEMS

Yijiang Peng, Qing Guo, Zhaofeng Zhang, and Yanyan Shan

The Key Laboratory of Urban Security and Disaster Engineering, Ministry of Education, Beijing University of Technology, Beijing 100124, China

ABSTRACT

The four-mid-node plane model of base force element method (BFEM) on complementary energy principle is used to analyze the rock mechanics problems. The method to simulate the crack propagation using the BFEM is proposed. And the calculation method of safety factor for rock mass stability was presented for the BFEM on complementary energy principle. The numerical researches show that the results of the BFEM are consistent with the results of conventional quadrilateral isoparametric element and quadrilateral reduced integration element, and the nonlinear BFEM has some advantages in dealing crack propagation and calculating safety factor of stability.

INTRODUCTION

The finite element method (FEM) has been playing a very important role in solving various problems in engineering and science. However, the conventional finite element method (FEM) based on the displacement model has some shortcomings, such as large deformation, treatment of incompressible materials, bending of thin plates, and moving boundary problems. In the past decades, numerous efforts techniques have been proposed for developing finite element models which are robust and insensitive to mesh distortion, such as the hybrid stress method [1–4], the equilibrium models [5, 6], the mixed approach [7], the integrated force method [8–11], the incompatible displacement modes [12, 13], the assumed strain method [14–17], the enhanced strain modes [18, 19], the selectively reduced integration scheme [20], the quasiconforming element method [21], the generalized conforming method [22], the Alpha finite element method [23], the new spline finite element method [24, 25], the unsymmetric method [26–29], the new natural coordinate methods [30–33], the smoothed finite element method [34], and the base force element method [35–43].

In recent years, some scholars are studying other types of numerical analysis methods, such as boundary element method [44, 45] and meshless method [46, 47]. And some scholars still adhere to explore the finite element method based on complementary energy principle [48–51]. However, these methods have not been widely applied in engineering.

In this paper, the base force element method (BFEM) on complementary energy principle is used to analyze the engineering problems of rock mechanics. The "base forces" was introduced by Gao [52], who used the concept to replace various stress tensors for the description of the stress state at a point. These base forces can be directly obtained from the strain energy. For large deformation problems, when the base forces were adopted, the derivation of basic formulae was simplified by Gao [53] and Gao et al. [54–56]. Based on the concept of the base forces, precise expressions for stiffness and compliance matrices for the FEM were obtained by Gao [52]. The applications of the stiffness matrix to the plane problems of elasticity using the plane quadrilateral element and the polygonal element were researched by Peng et al. [37]. Using the concept of base forces as state variables, a three-dimensional formulation of base force element method (BFEM) on complementary energy principle was proposed by Peng and Liu [35] for geometrically nonlinear problems. And the new finite element method based on the concept of base forces was called as the Base Force Element Method (BFEM) by Peng and Liu [35]. A three-dimensional model of base force element method (BFEM) on complementary energy principle was proposed by Liu and Peng [36] for elasticity problems. A 4-mid-node plane element model of the BFEM on complementary energy principle was proposed by Peng et al. [38] for geometrically nonlinear problem, which is derived by assuming that the stress is uniformly distributed on each edges of a plane element. In the paper [39], an arbitrary convex polygonal element model of the BFEM on complementary energy principle was proposed for geometrically nonlinear problem. In the paper [43], a 4-mid-node plane model of BFEM on complementary energy principle was researched, and its computational performance was studied. The convex polygonal element model of BFEM on complementary energy principle was given by Peng et al. [40] for arbitrary mesh problems. In the paper [41], the concave polygonal element model of BFEM on complementary energy principle was proposed for the concave polygonal mesh problems. In the paper [42], the BFEM on potential energy principle was used to analyze recycled aggregate concrete (RAC) on mesolevel, in which the model of BFEM with triangular element was derived, and the simulation results of the BFEM agree with the test results of recycled aggregate concrete. In recently, the BFEM on damage mechanics has been used to analyze the compressive strength, the size effects of compressive strength, and fracture process of concrete at mesolevel, and the analysis method is the new way for investigating fracture mechanism and numerical simulation of mechanical properties for concrete.

The purpose of this paper is to survey the base forces element method on complementary energy principle for large-scale computing problems in rock engineering problems.

MODEL OF THE BFEM

Compliance Matrix

Consider a 4-mid-node plane element as shown in Figure 1; the compliance matrix of a base force element can be obtained as [43]

$$C_{IJ} = \frac{1+v}{EA}\left(r_{IJ}U - \frac{v}{1+v}r_I \otimes r_J\right), \quad (I, J = 1, 2, 3, 4) \tag{1}$$

in which E is Young's modulus, v is Poisson's ratio, A is the area of an element, U is the unit tensor, and r_{IJ} is the dot product of radius vectors r_I and r_J at points I and J.

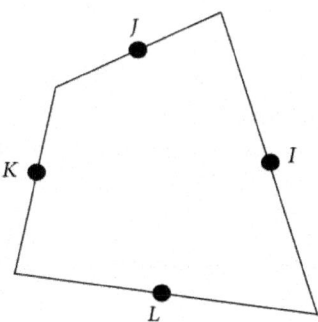

Figure 1: Four-mid-node plane element.

For a plane rectangular coordinate system, the radius vectors r_I and r_J of points I and J can be written as

$$\mathbf{r}_I = x_I \mathbf{e}_x + y_I \mathbf{e}_y, \qquad \mathbf{r}_J = x_J \mathbf{e}_x + y_J \mathbf{e}_y \tag{2}$$

in which e_x, e_y are the unit vectors.

Further, the compliance matrix of an element can be reduced as follows:

$$\begin{aligned}
C_{IJ} = \frac{1+v}{EA} \Bigg[&\left(\frac{1}{1+v} x_I x_J + y_I y_J \right) \mathbf{e}_x \otimes \mathbf{e}_x \\
&- \frac{v}{1+v} x_I y_J \mathbf{e}_x \otimes \mathbf{e}_y - \frac{v}{1+v} y_I x_J \mathbf{e}_y \otimes \mathbf{e}_x \\
&+ \left(x_I x_J + \frac{1}{1+v} y_I y_J \right) \mathbf{e}_y \otimes \mathbf{e}_y \Bigg].
\end{aligned} \tag{3}$$

For a plane strain problem, it is necessary to replace E by $E/(1-v^2)$ and v by $v/(1-v)$ in (1) and (3).

The characteristics of the BFEM on complementary energy principle are that the model does not introduce an interpolating function and is not necessary to introduce the Gauss integral for calculating the compliance coefficient at a point.

Governing Equations

The total complementary energy of the elastic system which has elements can be written as

$$\Pi_C = \sum_n \left(W_C^e - \overline{\mathbf{u}}_I \cdot \mathbf{T}^I \right), \tag{4}$$

where W_C^e is the complementary energy of an element T^I and $\overline{\mathbf{u}}_I$ are the resultant force vectors and the given displacement acting on the center node I of the edge I, respectively.

The equilibrium conditions can be released by the Lagrange multiplier method, and a new complementary energy function for an element can be introduced as follows:

$$\Pi_C^{e\,*} (\mathbf{T}, \lambda, \lambda_\theta) = \Pi_C^e (\mathbf{T}) + \lambda \left(\sum_{I=1}^{4} \mathbf{T}^I \right) + \lambda_\theta \left(\mathbf{T}^I \times \mathbf{r}_I \right), \tag{5}$$

where $\lambda = \lambda_x \mathbf{e}_x + \lambda_y \mathbf{e}_y$ and λ_θ are the Lagrange multipliers.

For the elastic system, the modified total complementary energy function of the elastic system which contains n elements should meet the following equation by means of the modified complementary energy principle:

$$\delta\Pi_C^* = \sum_n \left[\delta\Pi_C^{e\,*} \left(\mathbf{T}, \lambda, \lambda_\theta \right) \right] = 0. \tag{6}$$

Further, (6) can be expressed as

$$\frac{\partial\Pi_C^* \left(\mathbf{T}, \lambda, \lambda_\theta \right)}{\partial\mathbf{T}} = 0, \qquad \frac{\partial\Pi_C^* \left(\mathbf{T}, \lambda, \lambda_\theta \right)}{\partial\lambda} = 0,$$

$$\frac{\partial\Pi_C^* \left(\mathbf{T}, \lambda, \lambda_\theta \right)}{\partial\lambda_\theta} = 0. \tag{7}$$

The first of (7) is the compatibility equations and displacement boundary conditions for the elastic system. According to this equation, the displacement boundary conditions in this paper can be implemented in the BFEM. The second of (7) is the force equilibrium equation of each element. The third of (7) is the moment equilibrium equation of each element. These are the governing equations of the BFEM. From the equations, we can obtain the resultant forces acting on the center points of the edges of all elements.

Stress Tensor of an Element

Consider the 4-mid-node plane element as shown in Figure 1; the real stress σ of the element can be replaced by the average stress $\bar{\sigma}$ if the element is small enough. According to the definitions of Cauchy stress tensors, the stress expressions of an element can be obtained as

$$\sigma = \frac{1}{A} \mathbf{T}^I \otimes \mathbf{r}_I, \tag{8}$$

where $\otimes$ is the dyadic symbol, $\mathbf{T}^I$ and $\mathbf{r}_I$ are the resultant force vectors acting on the center node I of the edge I and the radius vector of the node I, respectively, and the summation rule is implied.

For a plane rectangular coordinate system, the force vectors $\mathbf{T}^I$ of the node I can be written as

$$\mathbf{T}^I = T_x^I \mathbf{e}_x + T_y^I \mathbf{e}_y, \tag{9}$$

where T_x^I and T_y^I are the components of the force vector $\mathbf{T}^I$ along coordinates x and y, respectively

Further, the stress tensors of an element can be reduced as follows:

$$\sigma = \frac{1}{A} \sum_{I=1}^{4} \left[T_x^I x_I \mathbf{e}_x \otimes \mathbf{e}_x + T_x^I y_I \mathbf{e}_x \otimes \mathbf{e}_y \right.$$

$$\left. + T_y^I x_I \mathbf{e}_y \otimes \mathbf{e}_x + T_y^I y_I \mathbf{e}_y \otimes \mathbf{e}_y \right]. \tag{10}$$

Displacement Vector of Nodes

According to the governing equation of an element of the BFEM, the explicit expression of displacement can be obtained as

$$\boldsymbol{\delta}_I = \mathbf{C}_{IJ} \cdot \mathbf{T}^J + \boldsymbol{\lambda} + \lambda_\theta \boldsymbol{\varepsilon} \cdot \mathbf{r}_I, \tag{11}$$

in which ε is the alternating tensor, λ and λ_θ are the Lagrange multipliers [43], and ε and λ can be expressed as

$$\boldsymbol{\varepsilon} = \mathbf{e}_x \otimes \mathbf{e}_y - \mathbf{e}_y \otimes \mathbf{e}_x,$$

$$\boldsymbol{\lambda} = \lambda_x \mathbf{e}_x + \lambda_y \mathbf{e}_y. \tag{12}$$

Further, the displacement vectors of an element can be reduced as follows [43]:

$$\boldsymbol{\delta}_I = \left(C_{IxJx} T_x^J + C_{IxJy} T_y^J + \lambda_x + \lambda_\theta y_I \right) \mathbf{e}_x$$

$$+ \left(C_{IyJx} T_x^J + C_{IyJy} T_y^J + \lambda_y - \lambda_\theta x_I \right) \mathbf{e}_y. \tag{13}$$

SIMULATION OF GRAVITY AND MATERIAL

In engineering problems, regardless of the dam, rock slope, or other structure of rock mass, the gravity of structure should be considered in the numerical calculation. We did not take into account the gravity problem when we previously prepared the program of BFEM on complementary energy principle. In order to consider the gravity of structure, three problems must be solved, including the calculation problem of equivalent node loads in the BFEM on complementary energy principle, the problem of exerting gravity in the software of the BFEM and the calculation problem of stress tensor in an element when the gravity is added.

Equivalent Node Loads of Gravity

For the BFEM on complementary energy principle, the equivalent node loads of gravity in an element can be calculated according to the principle of virtual work, as shown in Figure 2, and the expression can be given as

$$\{Q\}^e = -\frac{\rho g A t}{4}\begin{bmatrix} 0 & 1 & 0 & 1 & 0 & 1 & 0 & 1 \end{bmatrix}^T \tag{14}$$

or

$$\mathbf{Q}^I = -\frac{\rho g A t}{4}\mathbf{e}_y, \quad (I = 1, 2, 3, 4), \tag{15}$$

where t is the thickness of an element, ρ is the density of material, and g is acceleration of gravity.

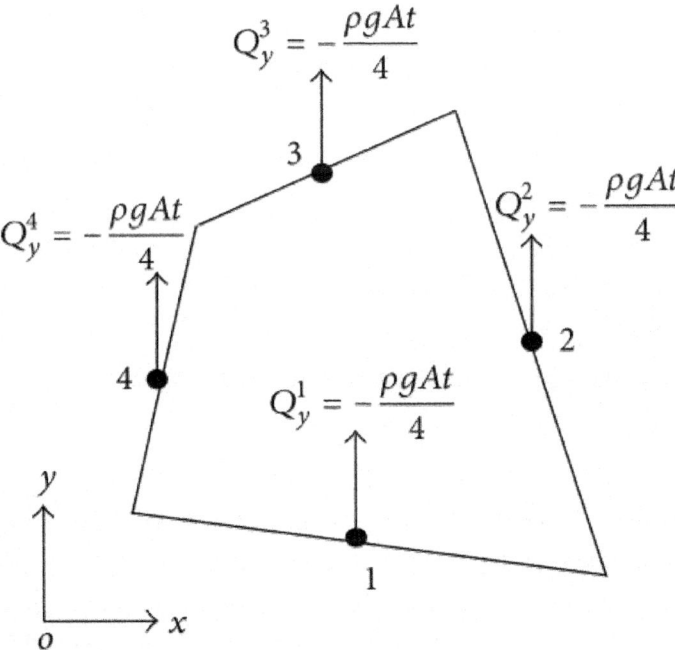

Figure 2: Equivalent node loads of gravity in an element.

Stress Calculation of an Element

When the gravity of an element is not taken into account, as shown in Figure 3, we calculate the force acting on the edges of an element first. Then, the stress tensor of the element can be calculated by (8) or (10).

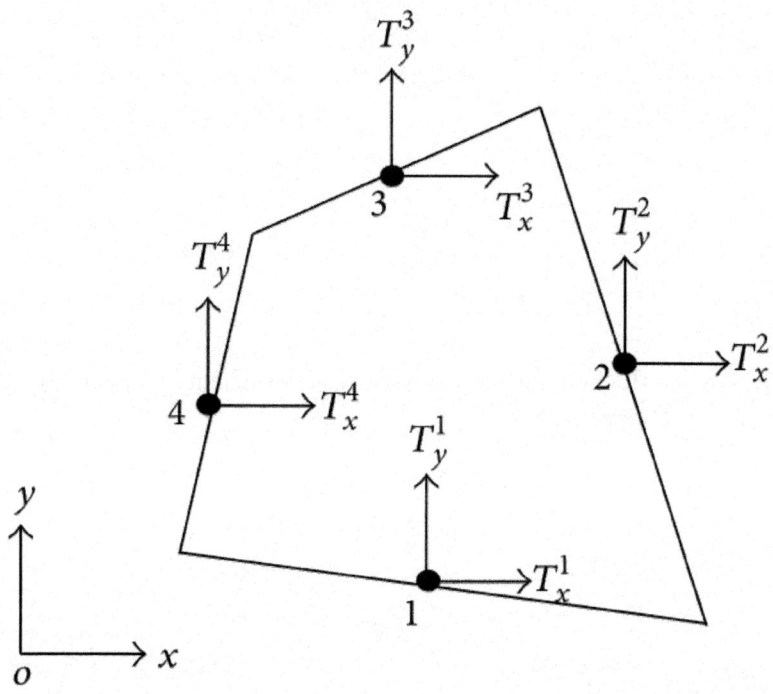

Figure 3: An element without gravity.

Further, the stress tensors of an element can also be reduced as follows:

$$\overline{\sigma} = \frac{1}{A}\left(\mathbf{T}^1 \otimes \mathbf{r}_1 + \mathbf{T}^2 \otimes \mathbf{r}_2 + \mathbf{T}^3 \otimes \mathbf{r}_3 + \mathbf{T}^4 \otimes \mathbf{r}_4\right),$$

(16)

where $\mathbf{T}^I = T_x^I \mathbf{e}_x + T_y^I \mathbf{e}_y$, $\mathbf{r}^I = x_I \mathbf{e}_x + y_I \mathbf{e}_y$, ($I = 1, 2, 3, 4$), x_I and y_I are the coordinates of node I, respectively.

When the gravity of an element is taken into account and there is a free boundary, as shown in Figure 4, the stress tensor of the element can be calculated as

$$\overline{\sigma} = \frac{1}{A}\Big[\left(\mathbf{T}^1 + \mathbf{Q}^1\right) \otimes \mathbf{r}_1 + \mathbf{Q}^2 \otimes \mathbf{r}_2$$

$$+ \left(\mathbf{T}^3 + \mathbf{Q}^3\right) \otimes \mathbf{r}_3 + \left(\mathbf{T}^4 + \mathbf{Q}^4\right) \otimes \mathbf{r}_4\Big].$$

(17)

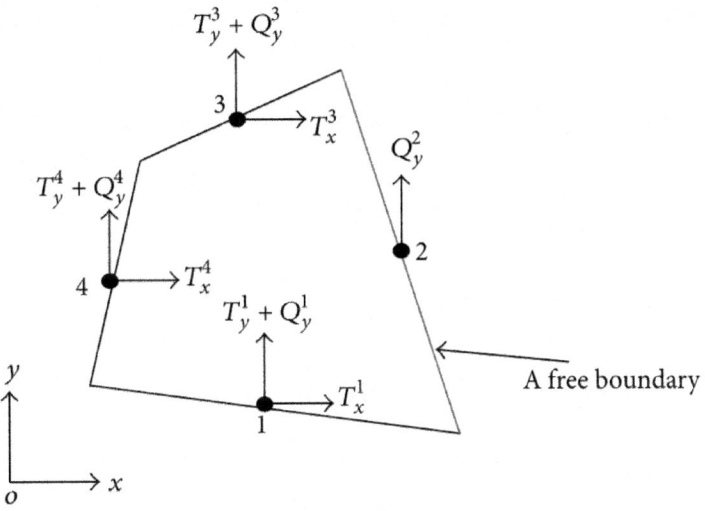

Figure 4: Considering gravity and a free boundary.

When the gravity of an element is taken into account and there is a force boundary condition, as shown in Figure 5, the stress tensor of the element can be calculated as

$$\overline{\sigma} = \frac{1}{A} \left[\left(\mathbf{T}^1 + \mathbf{Q}^1 \right) \otimes \mathbf{r}_1 + \left(\mathbf{F} + \mathbf{Q}^2 \right) \otimes \mathbf{r}_2 \right.$$

$$\left. + \left(\mathbf{T}^3 + \mathbf{Q}^3 \right) \otimes \mathbf{r}_3 + \left(\mathbf{T}^4 + \mathbf{Q}^4 \right) \otimes \mathbf{r}_4 \right] \tag{18}$$

in which $\mathbf{F} = F_x \mathbf{e}_x + F_y \mathbf{e}_y$.

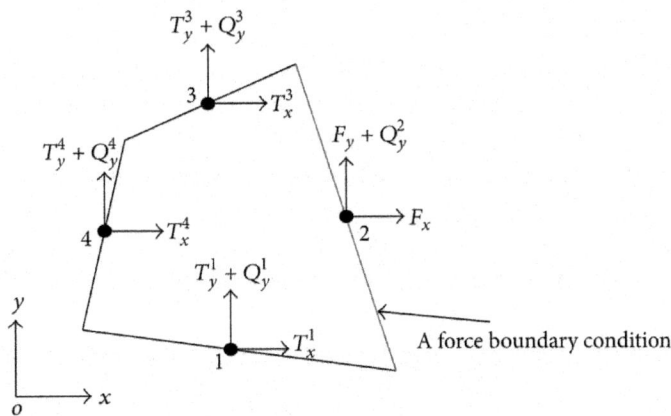

Figure 5: Considering gravity and a force boundary condition.

When the gravity of an element is not taken into account and there is a force boundary condition, as shown in Figure 6, the stress tensor of the element can be calculated as

$$\overline{\sigma} = \frac{1}{A}\left(\mathbf{T}^1 \otimes \mathbf{r}_1 + \mathbf{F} \otimes \mathbf{r}_2 + \mathbf{T}^3 \otimes \mathbf{r}_3 + \mathbf{T}^4 \otimes \mathbf{r}_4\right).$$

(19)

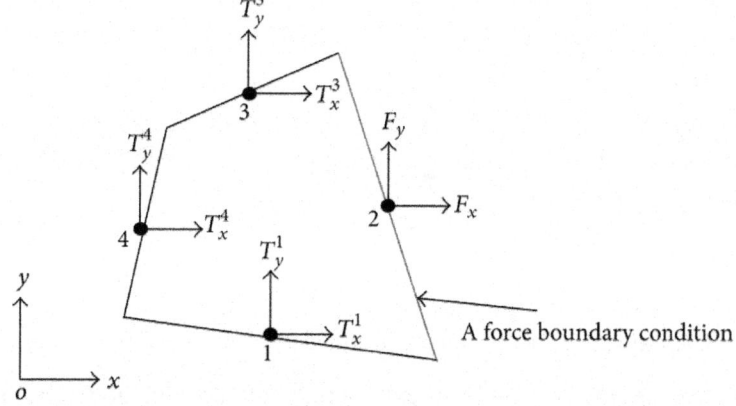

Figure 6: Considering a force boundary condition and no gravity.

Simulation of Different Materials

We adopt two one-dimensional array variables to reflect the change of elastic modulus and Poisson's ratio of different materials, respectively.

THE NONLINEAR MODEL OF BFEM FOR CRACK PROPAGATION PROBLEMS

The conventional displacement model of FEM requires the mesh reconstruction for the crack propagation problems. Therefore, the conventional FEM has deficiencies for the crack propagation problems. Because the contact forces between each element are used, the BFEM on complementary energy principle has advantage in the simulation of crack propagation problems. In the BFEM on complementary energy principle, we only need to deal with the constraint conditions and compatibility equations of displacement.

Failure Criteria of the Contact Interface between Two Elements

According to the control equations of the BFEM on complementary energy principle, the forces acting on the edge on an element can easily be calculated. According to the failure criteria expressed by the forces acting on the edge of an element, we can judge whether the interface of elements is cracking. If there is cracking of element interface, the nonlinear processing must be carried out.

Condition of Elastic State at Contact Interface

When $T_n < 0$ and $|T_s| < -f \cdot T_n + c \cdot l$ or $0 < T_n < t_0$ and $|T_s| < c \cdot l$, the contact interface is not cracked and in the elastic state. Here, T_n and T_s are the normal interface force and tangential interface force at contact interface between two elements, respectively. c and f are the cohesion and the internal friction coefficient of the contact interface, respectively. l is the length of an element. t_0 is the tensile strength of an element, and it is positive in tension.

Condition of Positive Slip at Contact Interface

When $T_n < 0$ and $T_s \geq -f \cdot T_n + c \cdot l$, the contact interface of the two elements began to slip. We need to get rid of the tangential displacement constraint at the contact interface between the two elements that it has been cracked and put the frictional force as load that is used to an initial force condition. The new load acting on the contact interface between the two elements in the second loop of calculation can be written as follows:

$$T_{s0} = f \cdot T_s. \qquad (20)$$

When $0 < T_n < t_0$ and $T_s \geq c \cdot l$, the contact interface of the two elements begins to slip. We need to get rid of the tangential displacement constraint and the normal displacement constraint at the contact interface of the two elements. And the contact interface will crack after slipping.

Condition of Negative Slip at Contact Interface

When $T_n < 0$ and $T_s \leq f \cdot T_n - c \cdot l$, the contact interface of the two elements began to slip. We need to get rid of the tangential displacement constraint at the contact interface between the two elements that it has been cracked, and put the frictional force as load that is an initial force condition. The new load acting on the contact interface between the two elements in the second loop of calculation is

$$T_{s0} = -f \cdot T_s. \qquad (21)$$

When $0 < T_n < t_0$ and $T_s \leq -c \cdot l$, the contact interface of the two elements begins to slip. We need to get rid of the tangential displacement constraint and the normal displacement constraint at the contact interfaces of elements. And the contact interface will crack after slipping.

Condition of Pull Cracking at Contact Interface

When $T_n \geq t_0$, the contact interface of the two elements begins to crack. We need to get rid of the tangential displacement constraint and the normal displacement constraint at the contact interfaces of elements.

After the above checks, the computer program of BFEM uses the new loads and constraint conditions to solve the governing equations of BFEM on complementary energy principle and obtains the new forces acting on the contact interfaces of elements. The program repeats the above checks until there are no new cracking at the contact interfaces or the solutions of nonlinear equations cannot be convergence since the interface cracks are too long.

Flow Chart of the Nonlinear BEFM for Crack Propagation Problems

The flow chart of the nonlinear BEFM of crack propagation problems can be shown in Figure 7.

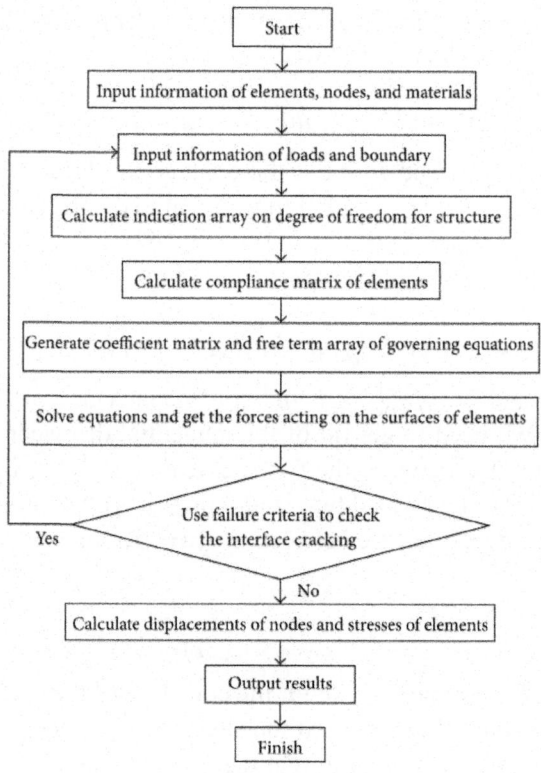

Figure 7: Flow chart of the nonlinear BEFM for crack propagation problems.

CALCULATION METHOD ON SAFETY FACTOR OF STABILITY IN THE BFEM

There are many methods to calculate the safety factor in engineering. The traditional finite element method usually used the rock joint elements to calculate factor of safety along the joint path. When the base force element method is used to analyze the stability of the rock mass in order to get the safety factor along the joint path, it is very easy. First, we calculate the surface forces of all elements according to the different load combinations. Then, we accumulate the sliding resistances and the sliding forces along the sliding path, respectively. Further, the safety factor of stability along the sliding path can obtained by the following equation:

$$K = \frac{\sum_{i=1}^{n} \left(-T_{ni} f_i + c_i l_i \right)}{\sum_{i=1}^{n} T_{si}},$$

(22)

where T_{ni} and T_{si} are the normal interface force and tangential interface force at contact interface between two elements along the sliding path, respectively. c_i and f_i are the cohesion and the internal friction coefficient of the contact interface between two elements along the sliding path, respectively. l_i is the length of the interface in the element along the sliding path.

NUMERICAL EXAMPLES

Example 1: A Rock Pillar under the Action of Gravity

Consider a thick pillar of rock under the action of gravity shown in Figure 8. And its width is 5 m, its height is 10 m, modulus of elasticity $E = 1 \times 10^8$ Pa, Poisson ratio $v = 0.3$, density of rock $\rho = 2.45$ t/m^3, and acceleration of gravity $g = 9.8$ m/s^2. The calculation is considered into the plane stress problem.

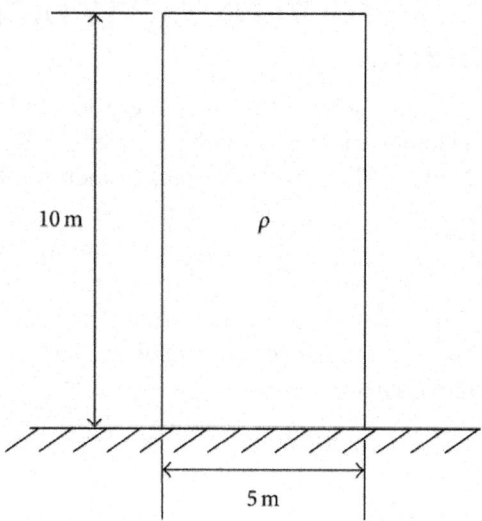

Figure 8: A rock pillar under the action of gravity.

The calculation is done using three different element meshes with the center nodes of edges of elements as shown in Figure 9.

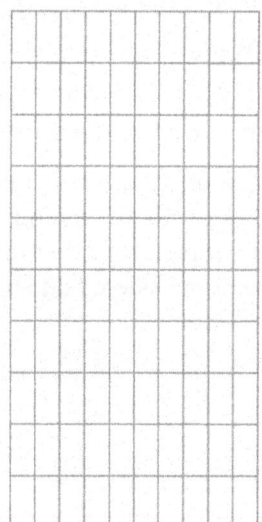

(a) 100 elements, 220 nodes

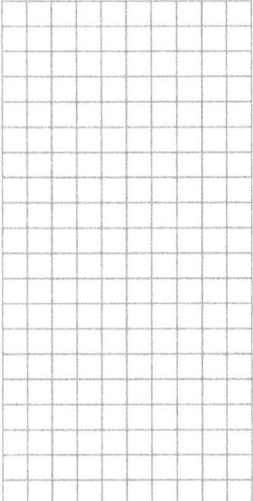

(b) 200 elements, 430 nodes

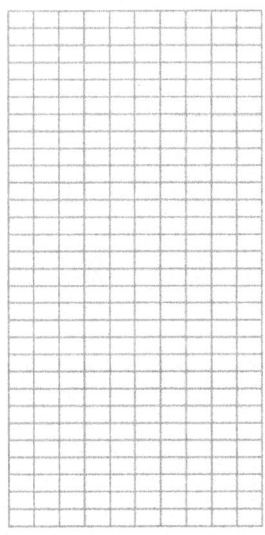

(c) 300 elements, 640 nodes

Figure 9: Three kinds of meshes for a rock pillar.

The values of stress components and displacement components of the rock pillar are listed in Tables 1–4, respectively. Comparisons of the results from the conventional quadrilateral isoparametric element (Q4 model) and quadrilateral reduced integration element (Q4R model) are also given in Tables 1–4, respectively. The numerical results of the present model are consistent with those of the Q4 model and Q4R model and have shown good computational stability.

Table 1: Displacement u_y at the top of the pillar

Meshes	BFEM model ($\times 10^{-2}$ m)	Q4 model ($\times 10^{-2}$ m)	Q4R model ($\times 10^{-2}$ m)
10 × 10	−1.1939	−1.1917	−1.2048
10 × 20	−1.1940	−1.1931	−1.1964
10 × 30	−1.1940	−1.1934	−1.1940

Table 2: Displacement u_y at $h=5$ m of the pillar

Meshes	BFEM ($\times 10^{-3}$ m)	Q4 model ($\times 10^{-3}$ m)	Q4R model ($\times 10^{-3}$ m)
10 × 10	−8.7592	−8.6975	−8.8045
10 × 20	−8.7630	−8.7118	−8.6726
10 × 30	−8.7637	−8.7149	−8.7363

Table 3: Stress σ_x at center of the elements at top of the pillar.

Meshes	BFEM (kPa)	Q4 model (kPa)	Q4R model (kPa)
10 × 10	−12.016	−12.244	−12.062
10 × 20	−6.005	−6.0773	−6.013
10 × 30	−4.003	−4.0382	−4.002

Table 4: Stress σ_y at center of the elements at bottom of the pillar.

Meshes	BFEM (kPa)	Q4 model (kPa)	Q4R model (kPa)
10 × 10	−244.194	−242.945	−244.349
10 × 20	−262.976	−260.116	−263.387
10 × 30	−272.102	−268.002	−272.552

Example 2: Analysis for a Rock Pillar with Four Materials under Pure Shear Effect

Consider a rock pillar with four materials under pure shear effect shown in Figure 10. And its modulus of elasticity $E_1 = 10^9$, $E_2 = 10^8$, $E_3 = 10^7$, and $E_4 = 10^6$, respectively. And Poisson ratio $v = 0.3$, the shear stress on surface of the structure $\tau=1$. The calculation is considered into the plane stress problem and the dimensionless values.

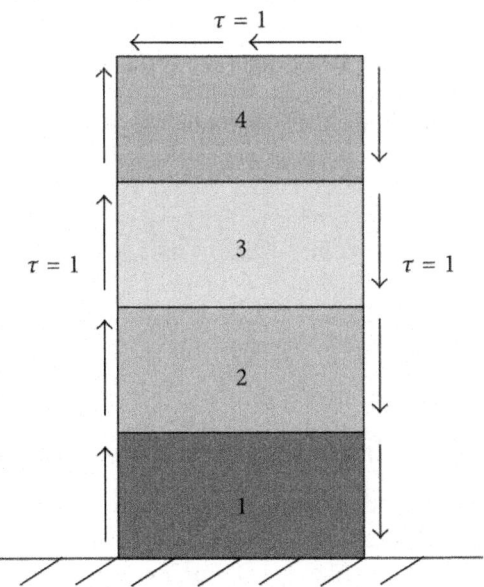

Figure 10: A rock pillar with four kinds of materials.

The calculation is done using the mesh with the center nodes of edges of elements as shown in Figure 11.

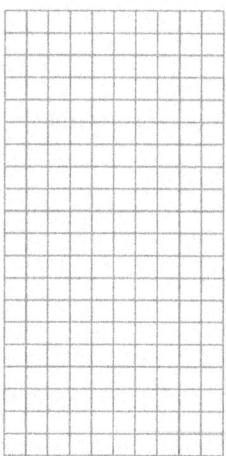

Figure 11: Mesh with 200 elements, 430 nodes.

The values of stress components of the rock pillar are listed in Table 5, respectively. Comparisons of the results from the theoretical analysis are also given in Table 5, respectively. The numerical results of the present model are consistent with those of the theoretical analysis.

Table 5: Stress solution of the pillar with four kinds of materials under uniform shearing forces.

	σ_x	τ_{xy}	σ_y
BFEM	0.000	1.000	0.000
Exact	0.000	1.000	0.000

Example 3: A Rock Pillar under Water Pressure and Gravity

Consider a rock pillar under water pressure and gravity shown in Figure 12. And its modulus of elasticity $E = 10^8$ Pa, Poisson' ratio $v = 0.3$, density of rock $\rho_1 = 2.4$ t/m³, density of water $\rho_2 = 1.0$ t/m³, and acceleration of gravity $g = 9.8$ m/s². The calculation is considered into the plane strain problem.

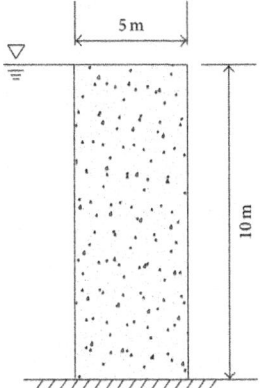

Figure 12: A rock pillar subjected to water pressure and gravity.

The calculation is done using four different element meshes with the center nodes of edges of elements as shown in Figure 13.

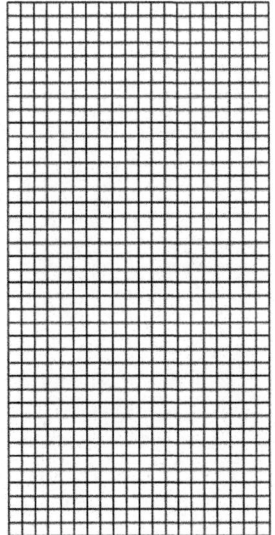

(a) 800 elements, 1660 nodes

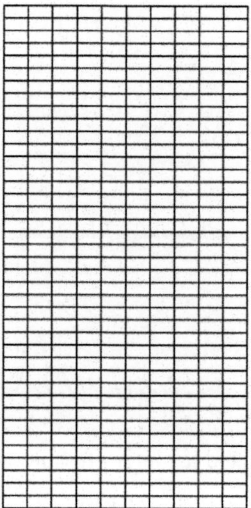

(b) 400 elements, 850 nodes

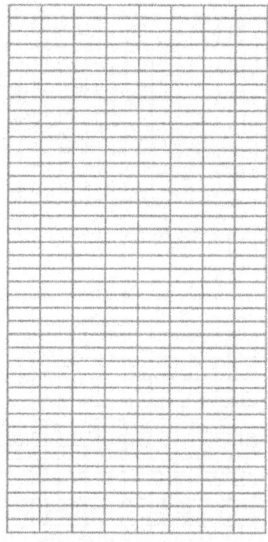

(c) 320 elements, 688 nodes

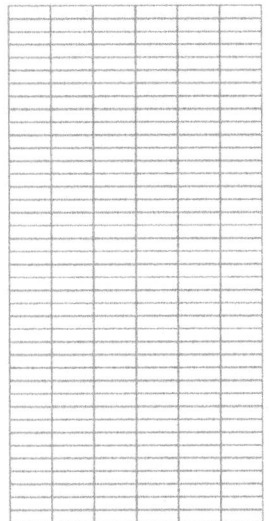

(d) 240 elements, 526 nodes

Figure 13: Four kinds of meshes for a rock pillar.

The values of stress components and displacement components of the rock pillar are listed in Tables 6–12, respectively. Comparisons of the results from the conventional quadrilateral isoparametric element (Q4 model) and quadrilateral reduced integration element (Q4R model) are also given in Tables 6–12, respectively. The numerical results of the present model are consistent with those of Q4R model, and the 4-mid-node element of BFEM has given good performance compared with Q4 model for the large aspect ratio of elements. Due to the different method calculating the equivalent node loads of water pressure between the base force elements and the element of traditional FEM, there is a slight error about the calculation results of the horizontal stresses σ_x of the element in lower right corner of rock pillar between the BFEM and the Q4R model, as shown in Table 10.

Table 6: Displacement u_x at $x = 2.5$ m of the pillar top.

Meshes	BFEM ($\times 10^{-2}$ m)	Q4 model ($\times 10^{-2}$ m)	Q4R model ($\times 10^{-2}$ m)
20×40	4.0219	4.0056	4.0190
10×40	4.0415	4.0094	4.0398
8×40	4.0563	4.0135	4.0532
6×40	4.0891	4.0239	4.0863

Table 7: Displacement u_x at h=5 m on the right of the pillar

Meshes	BFEM ($\times10^{-2}$ m)	Q4 model ($\times10^{-2}$ m)	Q4R model ($\times10^{-2}$ m)
20 × 40	2.0837	2.0758	2.0830
10 × 40	2.0816	2.0667	2.0815
8 × 40	2.0820	2.0626	2.0823
6 × 40	2.0849	2.0566	2.0862

Table 8: Displacement u_y at x = 2.5 m of the pillar top.

Meshes	BFEM ($\times10^{-3}$ m)	Q4 model ($\times10^{-3}$ m)	Q4R model ($\times10^{-3}$ m)
20 × 40	−9.8891	−9.8869	−9.8895
10 × 40	−9.8888	−9.8856	−9.8898
8 × 40	−9.8889	−9.8850	−9.8943
6 × 40	−9.8889	−9.8840	−9.8926

Table 9: Displacement u_y at h=5 m on the right of the pillar

Meshes	BFEM model ($\times10^{-2}$ m)	Q4 model ($\times10^{-2}$ m)	Q4R model ($\times10^{-2}$ m)
20 × 40	−1.5918	−1.5878	−1.5913
10 × 40	−1.5488	−1.5412	−1.5483
8 × 40	−1.5290	−1.51894	−1.5284
6 × 40	−1.4983	−1.4832	−1.4975

Table 10: Stress σ_x at center of the element at lower right of pillar

Meshes	BFEM (kPa)	Q4 model (kPa)	Q4R model (kPa)
20 × 40	−128.963	−179.706	−132.974
10 × 40	−142.729	−191.962	−150.839
8 × 40	−141.996	−193.919	−152.815
6 × 40	−134.954	−194.249	−150.941

Table 11: Stress τ_{xy} at center of the element at lower right of pillar

Meshes	BFEM (kPa)	Q4 model (kPa)	Q4R model (kPa)
20 × 40	152.442	172.480	153.366
10 × 40	153.439	167.708	154.637
8 × 40	151.262	163.966	152.612
6 × 40	146.664	157.624	148.288

Table 12: Stress σ_y at center of the element at lower right of pillar.

Meshes	BFEM (kPa)	Q4 model (kPa)	Q4R model (kPa)
20 × 40	−802.328	−796.173	−802.741
10 × 40	−707.932	−700.016	−708.582
8 × 40	−675.867	−666.975	−676.616
6 × 40	−634.179	−623.753	−635.039

Example 4: Stress Analysis of Concrete Gravity Dam

Consider a concrete gravity dam shown in Figure 14. And height of the dam is 65 m, bottom width is 49 m, the water level is 60 m, the elastic modulus of concrete E_1 = 15 GPa, Poisson ratio of concrete v_1 = 0.2, the elastic modulus of rock E_2 = 30 GPa, Poisson ratio of rock v_2 = 0.3, density of concrete is 2.45 t/m³, density of water is 1 t/m³, and acceleration of gravity g = 9.8 m/s². The calculation is considered into the plane strain problem and considered the effect of rock foundation of the dam.

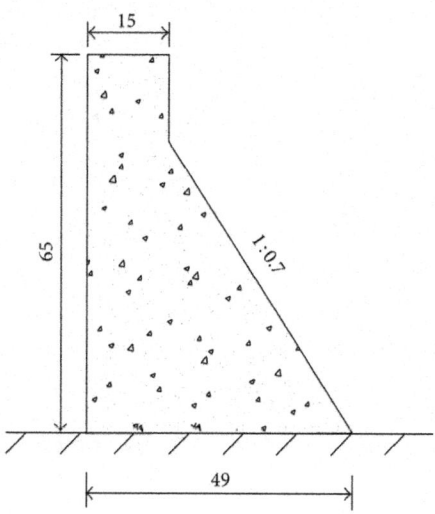

Figure 14: A concrete gravity dam.

The calculation is done using the mesh with the center nodes of edges of elements as shown in Figure 15. In this calculation, we do not consider the initial geostress field. The boundaries of the foundation are used the fixed constraint. The origin of coordinates is located at the bottom of the dam, and is 25 meters away from the dam heel.

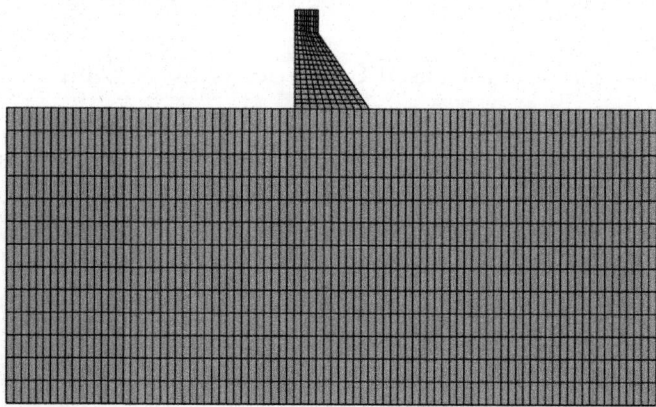

Figure 15: Meshes of the dam and its foundation (1344 elements, 2809 nodes).

The values of stress components and displacement components of the dam are plotted in Figures 16–25, respectively. Comparisons of the results from the conventional quadrilateral isoparametric element (Q4 model) and quadrilateral reduced integration element (Q4R model) are also given in Figures 16–25, respectively. The numerical results of the present model are consistent with those of the Q4 model and Q4R model and have shown good computational stability.

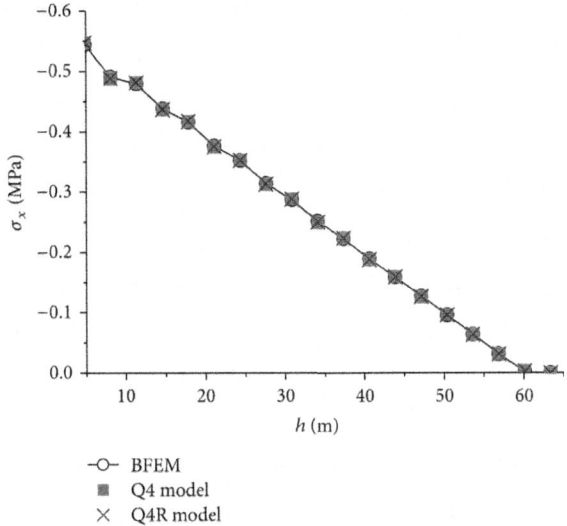

Figure 16: Figure 16: The h-σ_x curves at the upstream face of the dam.

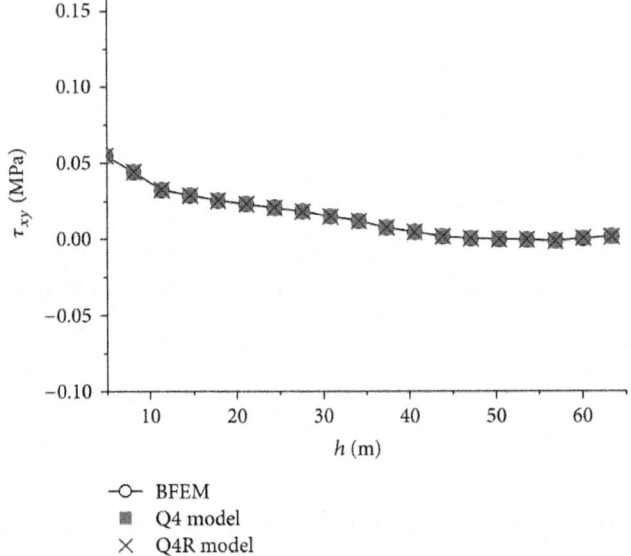

Figure 17: Figure 17: The h-τ_{xy} curves at the upstream face of the dam

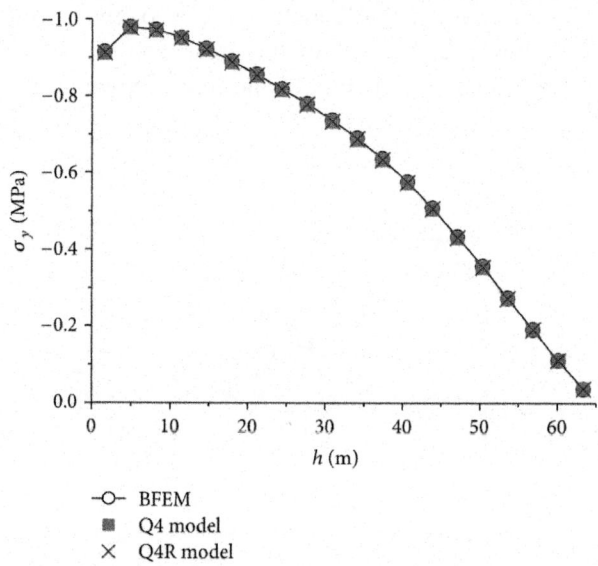

Figure 18: The h-σ_y curves at the upstream face of the dam.

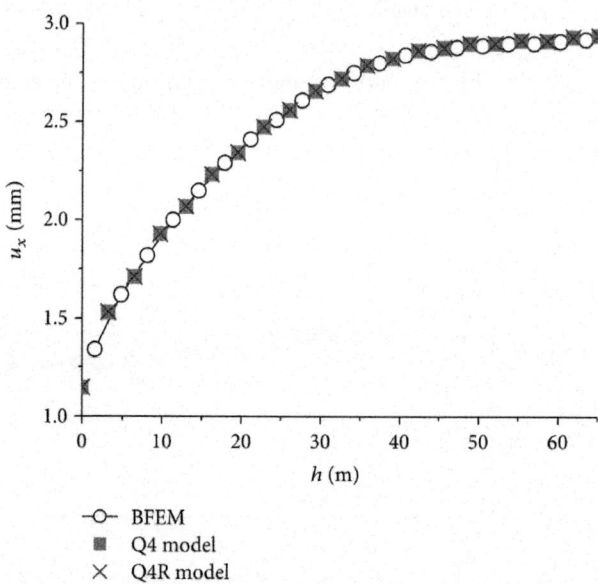

Figure 19: The h-u_x curves at the upstream face of the dam.

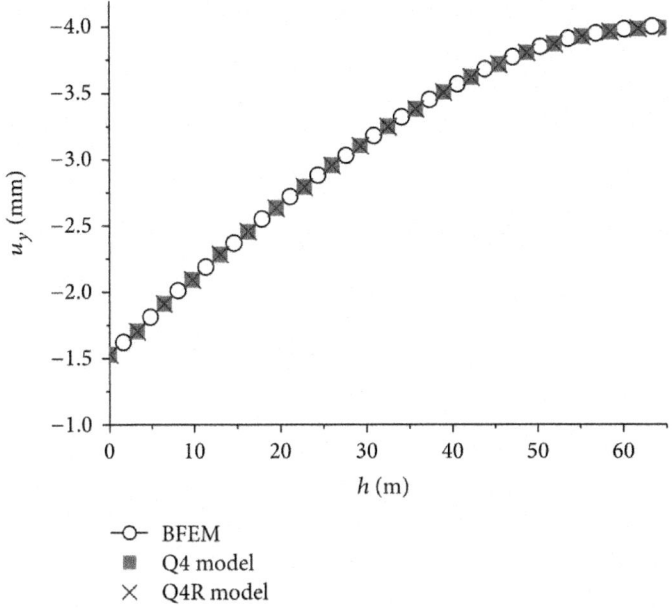

Figure 20: The h-u_y curves at the upstream face of the dam.

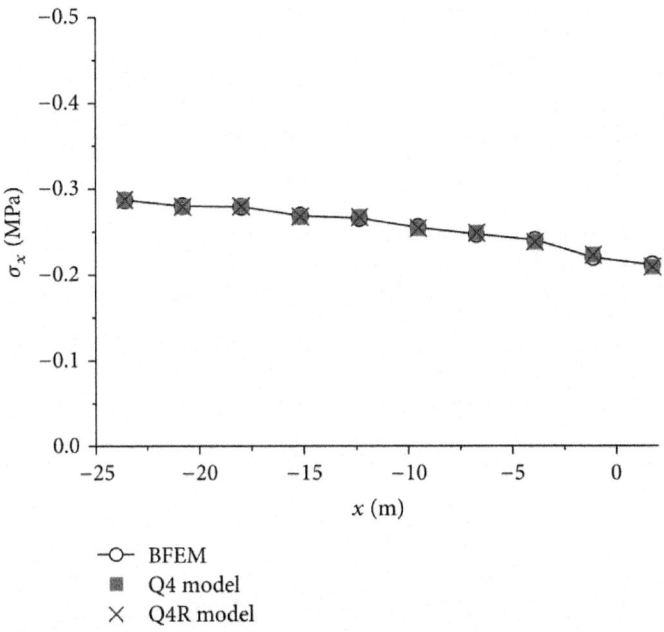

Figure 21: The h-σ_x curves at the half height of the dam.

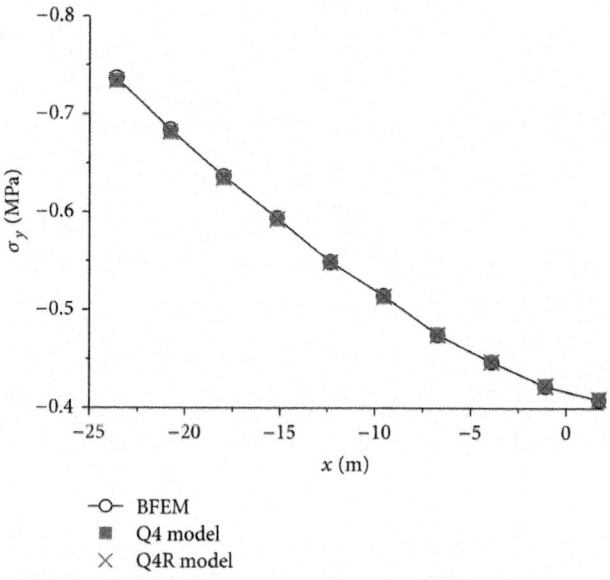

Figure 22: The h-σ_y curves at the half height of the dam

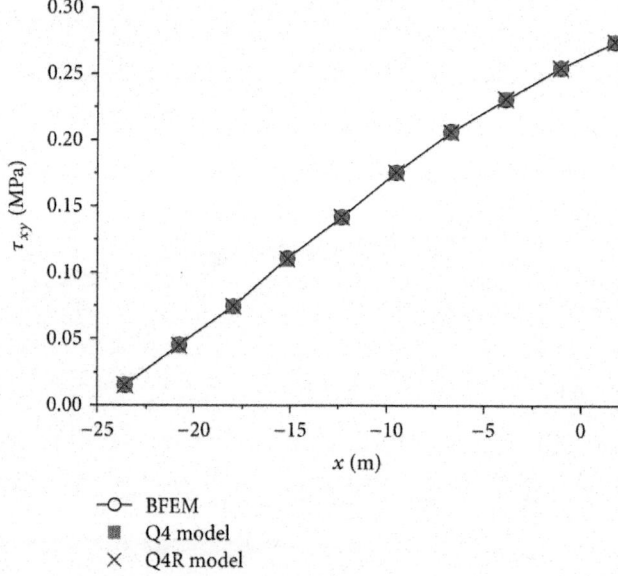

Figure 23: The h-τ_{xy} curves at the half height of the dam.

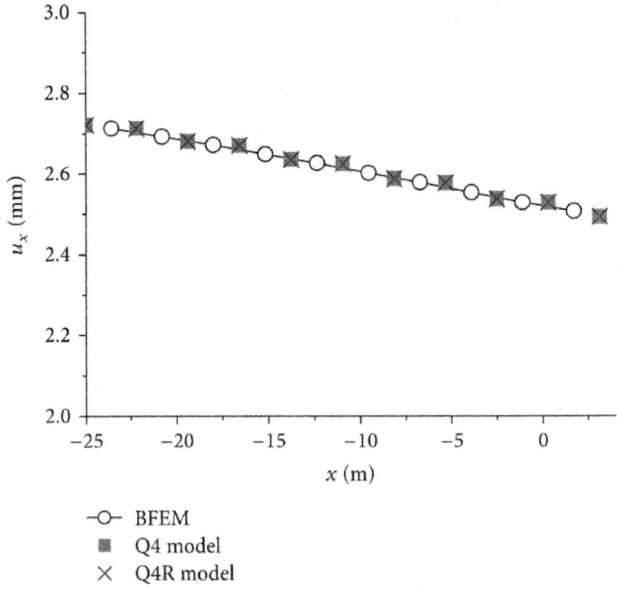

Figure 24: The h-u_x curves at the half height of the dam

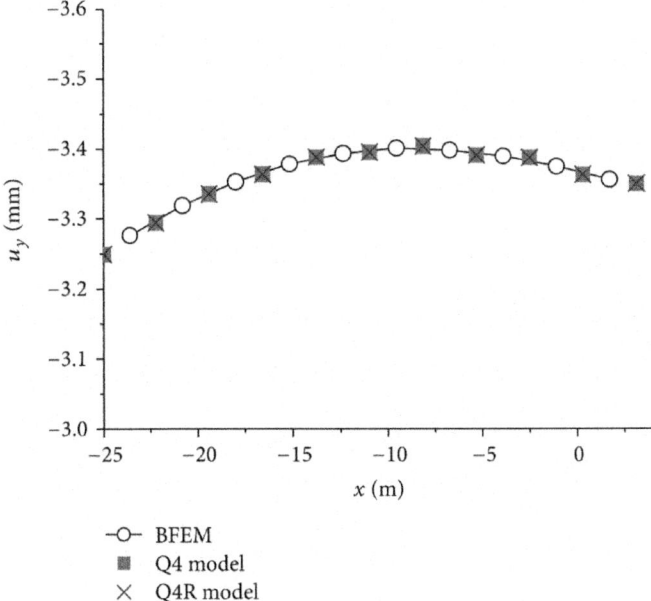

Figure 25: The h-u_y curves at the half height of the dam.

Example 5: Simulation and Analysis on the Horizontal Crack Propagation of Rock Block

Consider a rock block subjected by the horizontal thrust and vertical pressure shown in Figure 26. For the convenience of study, we do not consider the weight. And we use the dimensionless numerical analysis. Assuming elastic modulus $E=1$, Poisson ratio $v = 0.3$, tensile strength of a large number, and the uniform load $p_1 = 1$ and $p_2 = 1$. The calculation is considered into the plane stress problem and the dimensionless values.

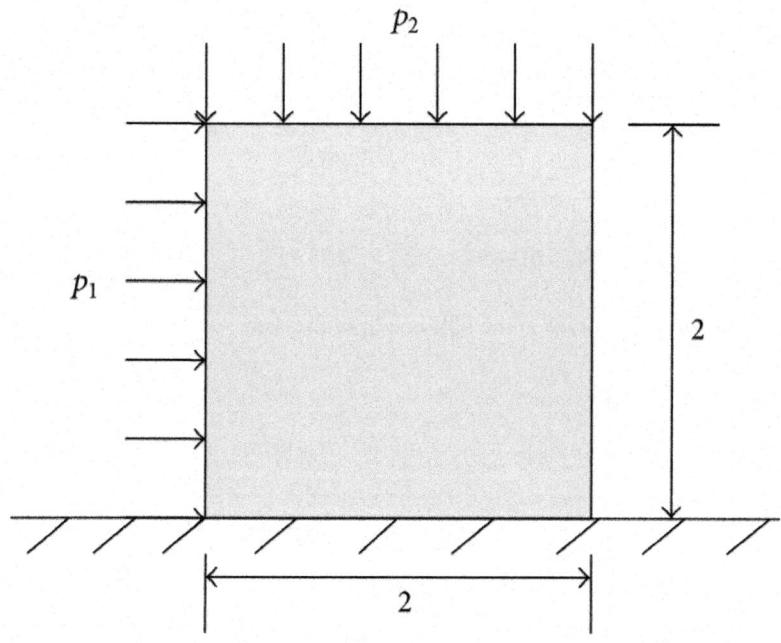

Figure 26: A rock block subjected by the horizontal thrust and vertical pressure.

The calculation is done using the mesh with the center nodes of edges of elements as shown in Figure 27.

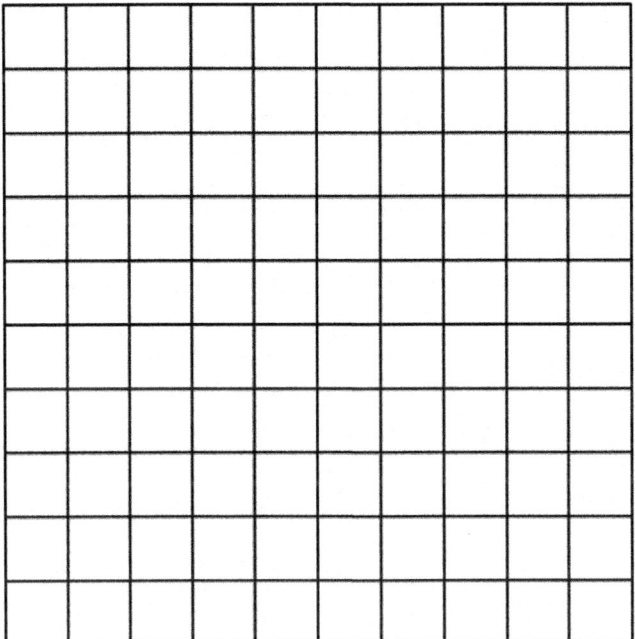

Figure 27: Meshes of the rock block.

In order to check whether the interface between the rock block and the ground will crack, we assume the friction coefficient of interface $f = 0.5$ and change the value of interface cohesion c. The results of calculation using the computer program of the nonlinear BFEM shown in Section 4 and the failure criteria in Section 4.1are as follows.

(1) When $c = 10$, there is no cracks.

(2) When $c = 2.8$, there is one element interface crack.

(3) When $c=2$, there are three elements' interfaces cracks.

(4) When $c = 1.5$, there are five elements' interfaces cracks.

(5) When $c = 1.1$, there are seven elements' interfaces cracks.

(6) When $c = 1.0$, the cracks are too long, and too little structural constraints have been insufficient to solve the equations

When the interface cohesion c is 1.5, the case of crack propagation of rock block is shown in Figure 28, and the safety factor of stability is $K=2$ which is consistent with the result of the theoretical analysis using the rigid limit equilibrium method.

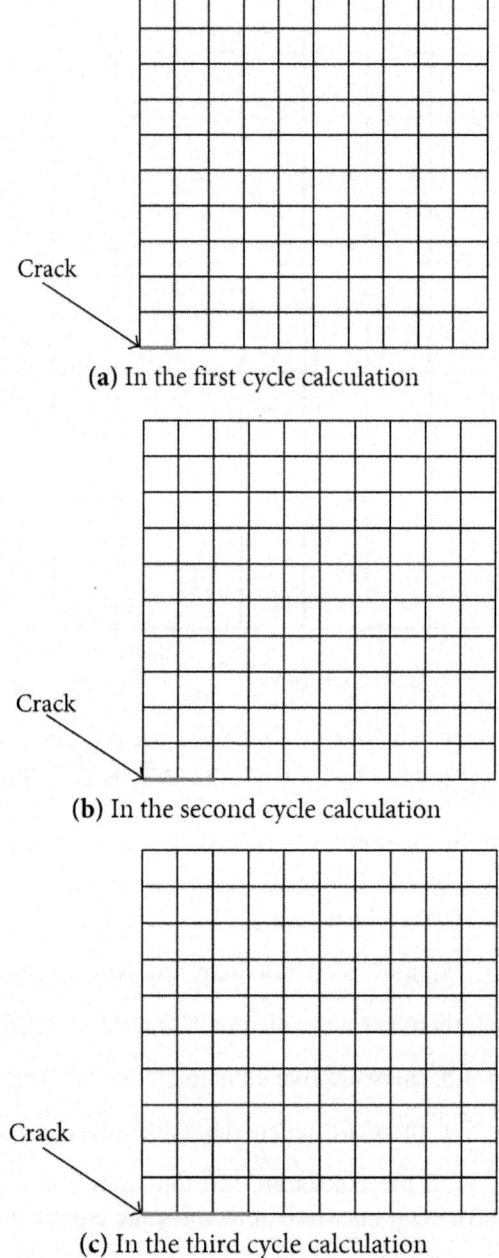

(a) In the first cycle calculation

(b) In the second cycle calculation

(c) In the third cycle calculation

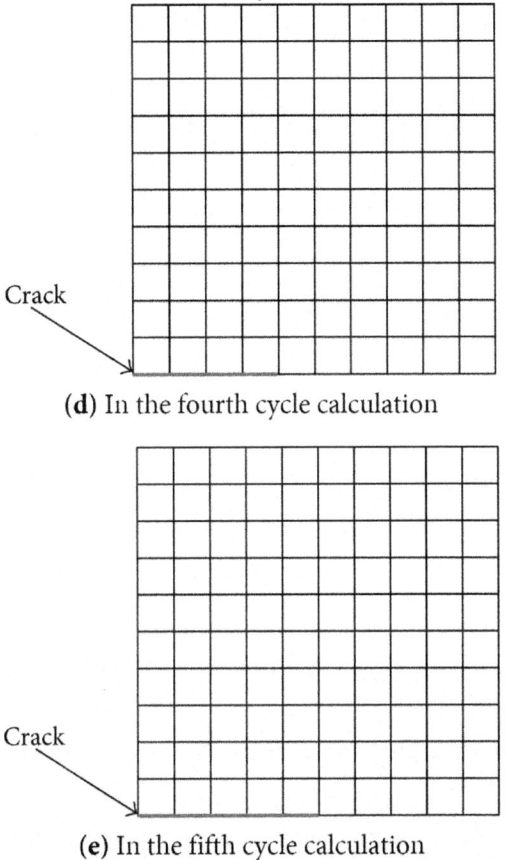

(d) In the fourth cycle calculation

(e) In the fifth cycle calculation

Figure 28: Crack propagation path of the rock block.

Example 6: Analysis on the Crack Propagation of Concrete Gravity Dam

Consider a concrete gravity dam shown in Figure 14. And height of the dam is 65 m, bottom width is 49 m, the water level is 60 m, The elastic modulus of concrete $E = 15$ GPa, Poisson ratio of concrete $v = 0.2$, density of concrete is 2.45 t/m³, density of water is 1 t/m³, and acceleration of gravity $g = 9.8$m/ s². The calculation is considered into the plane strain problem and is not considered the effect of rock foundation of the dam.

The calculation is done using the mesh with the center nodes of edges of elements as shown in Figure 29.

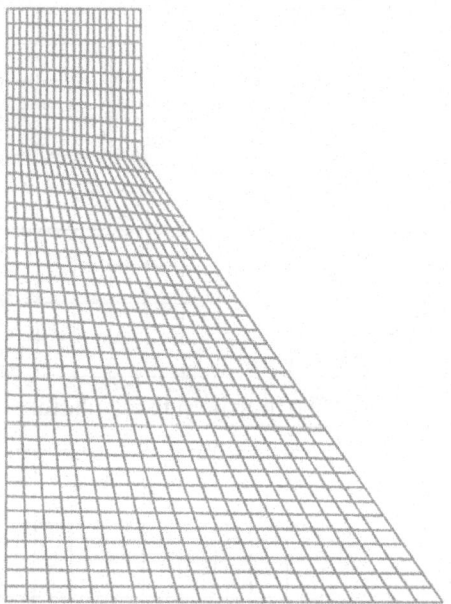

Figure 29: Mesh of gravity dam (800 elements, 1660 nodes).

No Initial Crack in Dam Foundation

We assume that the foundation surface of dam is weak structural interface and analyze the crack propagation and the safety factor by adjusting the value of the friction coefficient and the cohesion. The results of calculation using the computer program of the nonlinear BFEM shown in Section 4 and the failure criteria in Section 4.1 are as follows.

- When $c = 10^6$ Pa and $f = 0.95$, there is no cracks.
- When $c = 0.5 \times 10^6$ Pa and $f = 0.5$, there are two elements' interface cracks.
- When $c = 0.1 \times 10^6$ Pa and $f = 0.4$, there are eight elements' interfaces cracks.

The case of crack propagation of dam is shown in Figure 30. When $c = 10^6$ Pa and $f = 0.5$, the safety factor of stability $K = 4.0$ which is consistent with the result of the theoretical analysis. When $c = 0.5 \times 10^6$ Pa and $f = 0.5$, the safety factor of stability $K = 2.55$ which is consistent with the results of the theoretical analysis using the rigid limit equilibrium method.

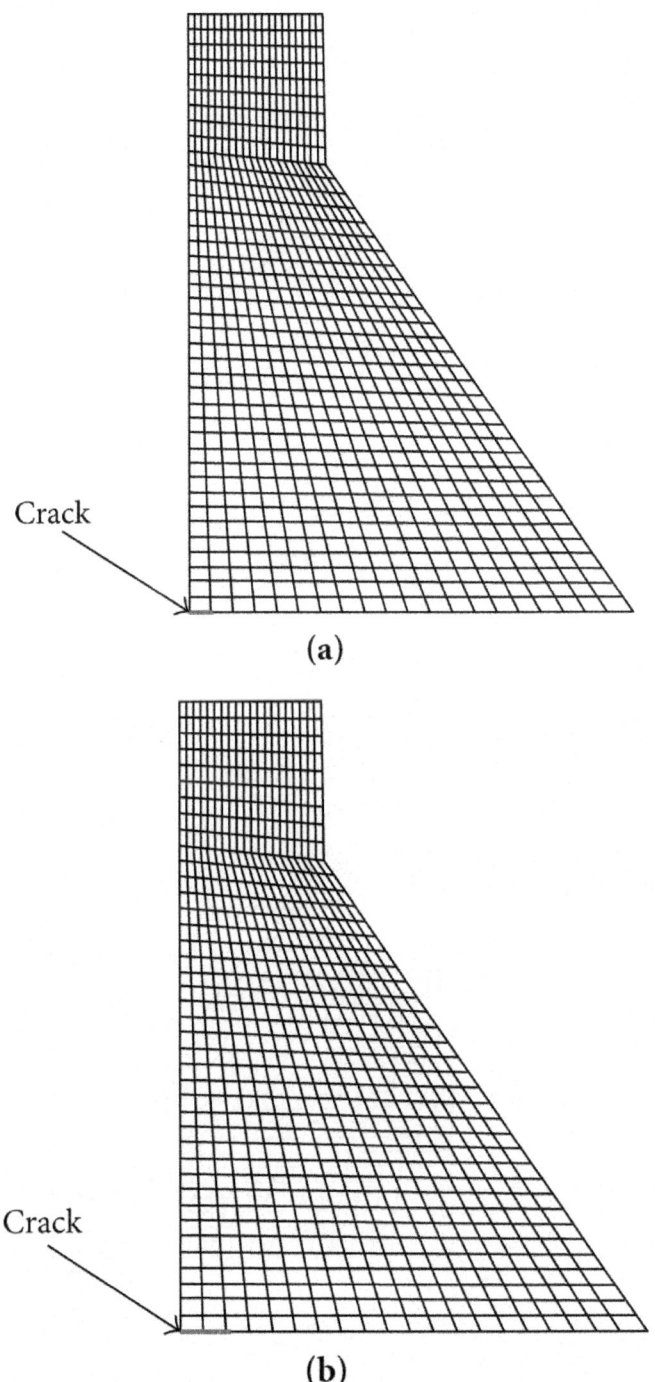

(a)

(b)

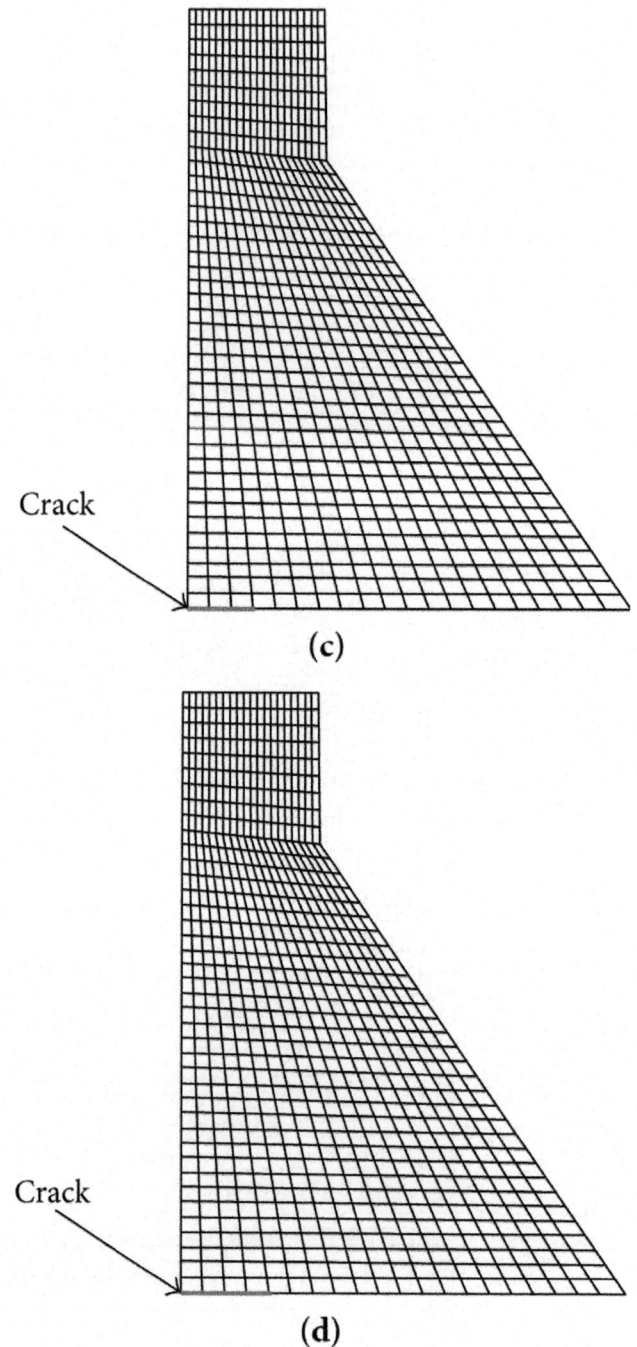

Crack

(c)

Crack

(d)

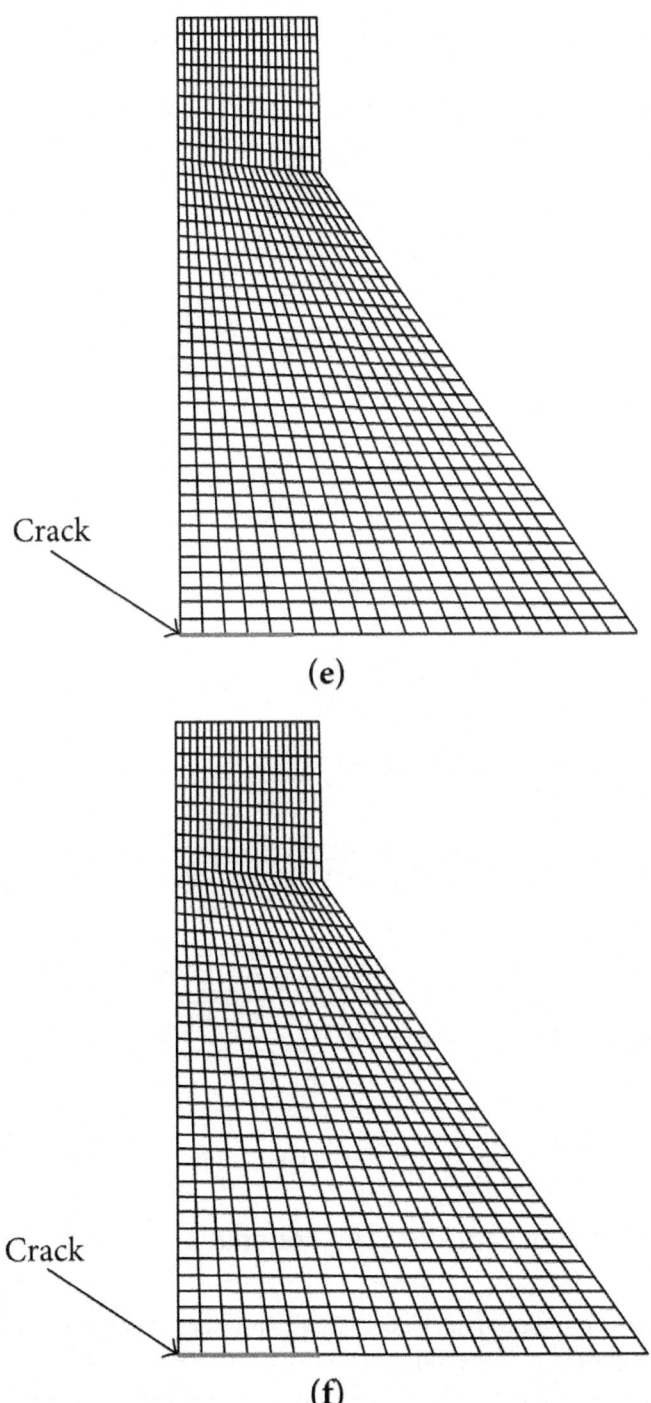

Crack

(e)

Crack

(f)

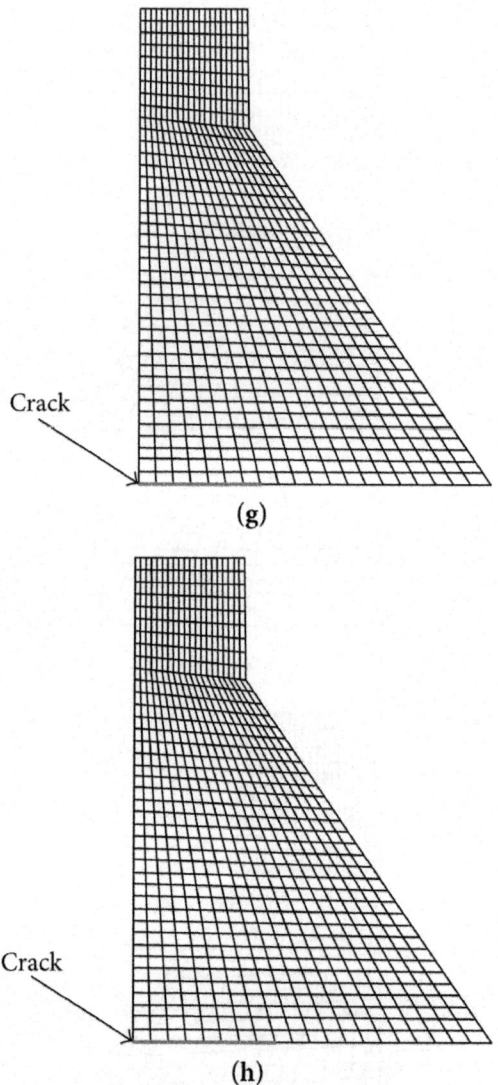

Figure 30: Crack propagation path of the gravity dam.

Existing an Initial Crack in Dam Foundation

There is an initial crack in the dam heel as shown in Figure 31. We assume that the dam foundation surface is weak structural interface and analyze the crack propagation and the safety factor by adjusting the value of the friction coefficient and the cohesion. The results of calculation using the computer program of the nonlinear BFEM shown in Section 4 and the failure criteria in

Section 4.1 are as follows.

(1) When $c = 10^6$ Pa and $f = 0.95$, there is no cracks.

(2) When $c = 0.5 \times 10^6$ Pa and $f = 0.5$, there is one element interface cracks.

(3) When $c = 0.1 \times 10^6$ Pa and $f = 0.5$, there are three element' interface cracks.

(4) When $c = 0.1 \times 10^6$ Pa and $f = 0.4$, there are seven elements' interfaces cracks.

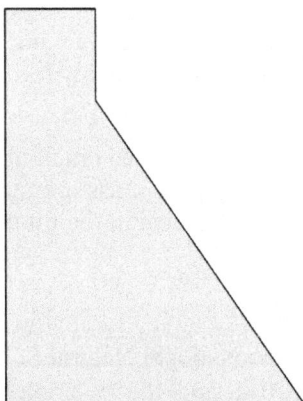

Figure 31: A gravity dam with a crack at the dam heel.

CONCLUSIONS

In this paper, the base force element method (BFEM) on complementary energy principle is used to analyze the rock mechanics problems. The methods to simulate the gravity of an element, the crack propagation and the safety factor of stability are proposed for the BFEM on complementary energy principle. The following conclusions can be drawn.

(1) The calculation results of the BFEM on complementary energy principle show that the numerical results of the present method coincide with the theoretical solution, the results of conventional quadrilateral isoparametric element (Q4 model) and quadrilateral reduced integration element (Q4R model). The correctness of the present method and its computer program is verified.

(2) The research results show that the BFEM on complementary energy principle has a good computational precision and stability is not sensitive to the effects on the aspect ratio of element and can be used for

large-scale scientific and engineering computing.

(3) The results of the BFEM for crack propagation problems show that the nonlinear BFEM can solve the cracking problem and simulate the crack propagation of the interface in rock mechanics engineering.

(4) The BFEM on complementary energy principle was applied to analyze the stability of rock mass and dam, and the results of safety factor are consistent with the results of the theoretical solutions using the rigid limit equilibrium method. The research results show the present method can be easily used to calculate the safety factor in rock engineering.

(5) This paper researched only the cracking problems with horizontal crack in rock mass and calculated only the safety factor of a single slip channel in rock mass.

(6) The cracking problems of inclined cracks and the safety factor of multiple sliding channels in rock mass are studying, and the further research results will be published in the future.

ACKNOWLEDGMENTS

This work is supported by the National Natural Science Foundation of China, nos. 10972015 and 11172015 and the preexploration project of the Key Laboratory of Urban Security and Disaster Engineering, Ministry of Education, Beijing University of Technology, no. USDE201404.

REFERENCES

1.   T. H. H. Pian, "Derivation of element stiffness matrices by assumed stress distributions," AIAA Journal, vol. 2, no. 7, pp. 1333–1336, 1964.

2.   T. H. Pian and D. P. Chen, "Alternative ways for formulation of hybrid stress elements," International Journal for Numerical Methods in Engineering, vol. 18, no. 11, pp. 1679–1684, 1982.

3.   T. H. H. Pian and K. Sumihara, "Rational approach for assumed stress finite elements," International Journal for Numerical Methods in Engineering, vol. 20, no. 9, pp. 1685–1695, 1984.

4.   C. Zhang, D. Wang, J. Zhang, W. Feng, and Q. Huang, "On the equivalence of various hybrid finite elements and a new orthogonalization method for explicit element stiffness formulation," Finite Elements in Analysis and Design, vol. 43, no. 4, pp. 321–332, 2007.

5.   B. Fraeijs de Veubeke, "Displacement and equilibrium models in the finite element method," in Stress Analysis, O. C. Zienkiewicz and G. S.

Holister, Eds., pp. 145–197, John Wiley & Sons, New York, NY, USA, 1965.

6.   B. F. de Veubeke, "A new variational principle for finite elastic displacements," International Journal of Engineering Science, vol. 10, no. 9, pp. 745–763, 1972.

7.   R. L. Taylor and O. C. Zienkiewicz, "Complementary energy with penalty functions in finite element analysis," in Energy Methods in Finite Element Analysis, R. Glowinski, Ed., pp. 153–174, John Wiley & Sons, New York, NY, USA, 1979.

8.   S. N. Patniak, "An integrated force method for discrete analysis," International Journal for Numerical Methods in Engineering, vol. 6, no. 2, pp. 237–251, 1973.

9.   S. N. Patnaik, "The integrated force method versus the standard force method," Computers and Structures, vol. 22, no. 2, pp. 151–163, 1986.

10. S. N. Patnaik, "The variational energy formulation for the integrated force method," AIAA Journal, vol. 24, no. 1, pp. 129–137, 1986.

11. S. N. Patnaik, L. Berke, and R. H. Gallagher, "Integrated force method versus displacement method for finite element analysis," Computers and Structures, vol. 38, no. 4, pp. 377–407, 1991.

12. E. L. Wilson, R. L. Tayler, W. P. Doherty, and J. Ghaboussi, "Incompatible displacement models," inNumerical and Computational Methods in Structural Mechanics, S. J. Fenves, N. Perrone, A. R. Robinson, and W. C. Schnobrich, Eds., pp. 43–57, Academic Press, New York, NY, USA, 1973.

13. R. L. Taylor, P. J. Beresford, and E. L. Wilson, "A non-conforming element for stress analysis,"International Journal for Numerical Methods in Engineering, vol. 10, no. 6, pp. 1211–1219, 1976.

14. J. C. Simo and T. J. R. Hughes, "On the variational foundations of assumed strain methods," Journal of Applied Mechanics, vol. 53, no. 1, pp. 51–54, 1986.

15. J. C. Simo and M. S. Rifai, "Class of mixed assumed strain methods and the method of incompatible modes," International Journal for Numerical Methods in Engineering, vol. 29, no. 8, pp. 1595–1638, 1990.

16. R. H. Macneal, "Derivation of element stiffness matrices by assumed strain distributions," Nuclear Engineering and Design, vol. 70, no. 1, pp. 3–12, 1982.

17. T. Belytschko and L. P. Bindeman, "Assumed strain stabilization of the 4-node quadrilateral with 1-point quadrature for nonlinear problems,"

Computer Methods in Applied Mechanics and Engineering, vol. 88, no. 3, pp. 311–340, 1991.

18. R. Piltner and R. L. Taylor, "A quadrilateral mixed finite element with two enhanced strain modes,"International Journal for Numerical Methods in Engineering, vol. 38, no. 11, pp. 1783–1808, 1995.

19. R. Piltner and R. L. Taylor, "A systematic constructions of B-bar functions for linear and nonlinear mixed-enhanced finite elements for plane elasticity problems," International Journal for Numerical Methods in Engineering, vol. 44, no. 5, pp. 615–639, 1997.

20. T. J. R. Hughes, "Generalization of selective integration procedures to anisotropic and nonlinear media,"International Journal for Numerical Methods in Engineering, vol. 15, no. 9, pp. 1413–1418, 1980.

21. T. Limin, C. Wanji, and L. Yingxi, "Formulation of quasi-conforming element and Hu-Washizu principle," Computers and Structures, vol. 19, no. 1-2, pp. 247–250, 1984.

22. L. Yu-qiu and H. Min-feng, "A generalized conforming isoparametric element," Applied Mathematics and Mechanics, vol. 9, no. 10, pp. 929–936, 1988.

23. G. R. Liu, T. Nguyen-Thoi, and K. Y. Lam, "A novel alpha finite element method (αFEM) for exact solution to mechanics problems using triangular and tetrahedral elements," Computer Methods in Applied Mechanics and Engineering, vol. 197, no. 45–48, pp. 3883–3897, 2008.

24. J. Chen, C.-J. Li, and W.-J. Chen, "A 17-node quadrilateral spline finite element using the triangular area coordinates," Applied Mathematics and Mechanics (English Edition), vol. 31, no. 1, pp. 125–134, 2010.

25. J. Chen, C.-J. Li, and W.-J. Chen, "A family of spline finite elements," Computers and Structures, vol. 88, no. 11-12, pp. 718–727, 2010.

26. S. Rajendran and K. M. Liew, "A novel unsymmetric 8-node plane element immune to mesh distortion under a quadratic displacement field," International Journal for Numerical Methods in Engineering, vol. 58, no. 11, pp. 1713–1748, 2003.

27. S. Rajendran, "A technique to develop mesh-distortion immune finite elements," Computer Methods in Applied Mechanics and Engineering, vol. 199, no. 17–20, pp. 1044–1063, 2010.

28. E. T. Ooi, S. Rajendran, and J. H. Yeo, "A 20-node hexahedron element with enhanced distortion tolerance," International Journal for Numerical Methods in Engineering, vol. 60, no. 15, pp. 2501–2530, 2004.

29. E. T. Ooi, S. Rajendran, and J. H. Yeo, "Remedies to rotational frame

dependence and interpolation failure of US-QUAD8 element," Communications in Numerical Methods in Engineering, vol. 24, no. 11, pp. 1203–1217, 2008.

30. Y. Long, L. Juxuan, Z. Long, and C. Song, "Area co-ordinates used in quadrilateral elements,"Communications in Numerical Methods in Engineering, vol. 15, no. 8, pp. 533–543, 1999.

31. Y. Q. Long, S. Cen, and Z. F. Long, Advanced Finite Element Method in Structural Engineering, Springer/Tsinghua University Press, Berlin, Germany, 2009.

32. Z. F. Long, J. X. Li, S. Cen, and Y. Q. Long, "Some basic formulae for area coordinates used in quadrilateral elements," Communications in Numerical Methods in Engineering, vol. 15, no. 12, pp. 841–852, 1999.

33. Z.-F. Long, S. Cen, L. Wang, X.-R. Fu, and Y.-Q. Long, "The third form of the quadrilateral area coordinate method (QACM-III): theory, application, and scheme of composite coordinate interpolation," Finite Elements in Analysis and Design, vol. 46, no. 10, pp. 805–818, 2010.

34. G. R. Liu, K. Y. Dai, and T. T. Nguyen, "A smoothed finite element method for mechanics problems,"Computational Mechanics, vol. 39, no. 6, pp. 859–877, 2007.

35. Y. Peng and Y. Liu, "Base force element method of complementary energy principle for large rotation problems," Acta Mechanica Sinica, vol. 25, no. 4, pp. 507–515, 2009.

36. Y. Liu and Y. Peng, "Base force element method (BFEM) on complementary energy principle for linear elasticity problem," Science China: Physics, Mechanics and Astronomy, vol. 54, no. 11, pp. 2025–2032, 2011.

37. Y. Peng, Z. Dong, B. Peng, and Y. Liu, "Base force element method (BFEM) on potential energy principle for elasticity problems," International Journal of Mechanics and Materials in Design, vol. 7, no. 3, pp. 245–251, 2011.

38. Y. Peng, Z. Dong, B. Peng, and N. Zong, "The application of 2D base force element method (BFEM) to geometrically non-linear analysis," International Journal of Non-Linear Mechanics, vol. 47, no. 3, pp. 153–161, 2012.

39. Y.-J. Peng, J.-W. Pu, B. Peng, and L.-J. Zhang, "Two-dimensional model of base force element method (BFEM) on complementary energy principle for geometrically nonlinear problems," Finite Elements in Analysis and Design, vol. 75, pp. 78–84, 2013.

40. Y. J. Peng, N. N. Zong, L. J. Zhang, and J. W. Pu, "Application of 2D base force element method with complementary energy principle for arbitrary meshes," Engineering Computations, vol. 31, no. 4, pp. 1–15, 2014.

41. Y. Peng, L. Zhang, J. Pu, and Q. Guo, "A two-dimensional base force element method using concave polygonal mesh," Engineering Analysis with Boundary Elements, vol. 42, pp. 45–50, 2014.

42. Y. Peng, Y. Liu, J. Pu, and L. Zhang, "Application of base force element method to mesomechanics analysis for recycled aggregate concrete," Mathematical Problems in Engineering, vol. 2013, Article ID 292801, 8 pages, 2013.

43. Y. Liu, Y. Peng, L. Zhang, and Q. Guo, "A 4-mid-node plane model of base force element method on complementary energy principle," Mathematical Problems in Engineering, vol. 2013, Article ID 706759, 8 pages, 2013.

44. C. Y. Dong and G. L. Zhang, "Boundary element analysis of three dimensional nanoscale inhomogeneities," International Journal of Solids and Structures, vol. 50, no. 1, pp. 201–208, 2013.

45. C. Y. Dong and E. Pan, "Boundary element analysis of nanoinhomogeneities of arbitrary shapes with surface and interface effects," Engineering Analysis with Boundary Elements, vol. 35, no. 8, pp. 996–1002, 2011.

46. S. S. Chen, Q. H. Li, Y. H. Liu, and Z. Q. Xue, "A meshless local natural neighbour interpolation method for analysis of two-dimensional piezoelectric structures," Engineering Analysis with Boundary Elements, vol. 37, no. 2, pp. 273–279, 2013.

47. S. Chen, Y. Liu, J. Li, and Z. Cen, "Performance of the MLPG method for static shakedown analysis for bounded kinematic hardening structures," European Journal of Mechanics, A/Solids, vol. 30, no. 2, pp. 183–194, 2011.

48. S. Cen, X.-R. Fu, and M.-J. Zhou, "8- and 12-node plane hybrid stress-function elements immune to severely distorted mesh containing elements with concave shapes," Computer Methods in Applied Mechanics and Engineering, vol. 200, no. 29–32, pp. 2321–2336, 2011.

49. S. Cen, G.-H. Zhou, and X.-R. Fu, "A shape-free 8-node plane element unsymmetric analytical trial function method," International Journal for Numerical Methods in Engineering, vol. 91, no. 2, pp. 158–185, 2012.

50. H. A. F. A. Santos, "Complementary-energy methods for geometrically non-linear structural models: an overview and recent developments in the analysis of frames," Archives of Computational Methods in Engineering, vol. 18, no. 4, pp. 405–440, 2011.

51. H. A. F. A. Santos and C. I. Almeida Paulo, "On a pure complementary energy principle and a force-based finite element formulation for nonlinear elastic cables," International Journal of Non-Linear Mechanics, vol. 46, no. 2, pp. 395–406, 2011.

52. Y. C. Gao, "A new description of the stress state at a point with applications," Archive of Applied Mechanics, vol. 73, no. 3-4, pp. 171–183, 2003.

53. Y. C. Gao, "Asymptotic analysis of the nonlinear Boussinesq problem for a kind of incompressible rubber material (compression case)," Journal of Elasticity, vol. 64, no. 2-3, pp. 111–130, 2001.

54. Y. C. Gao and T. J. Gao, "Large deformation contact of a rubber notch with a rigid wedge," International Journal of Solids and Structures, vol. 37, no. 32, pp. 4319–4334, 2000.

55. Y. C. Gao and S. H. Chen, "Analysis of a rubber cone tensioned by a concentrated force," Mechanics Research Communications, vol. 28, no. 1, pp. 49–54, 2001.

56. Y.-C. Gao, M. Jin, and G.-S. Dui, "Stresses, singularities, and a complementary energy principle for large strain elasticity," Applied Mechanics Reviews, vol. 61, no. 3, Article ID 030801, 16 pages, 2008.

Chapter 8

ROCK MAGNETIC PROPERTIES OF SEDIMEN-
TARY ROCKS IN CENTRAL HOKKAIDO —
INSIGHTS INTO SEDIMENTARY AND TECTON-
IC PROCESSES ON AN ACTIVE MARGIN

Yasuto Itoh[1], Machiko Tamaki[2], and Osamu Takano[3]

[1]Graduate School of Science, Osaka Prefecture University, Osaka, Japan

[2]Japan Oil Engineering Co. Ltd., Tokyo, Japan

[3]JAPEX Research Center, Japan Petroleum Exploration Co. Ltd., Chiba, Japan

INTRODUCTION

Reflecting a complicated subduction and collision history on the eastern Eurasian margin, central Hokkaido has been a site of various types of basin formation. Thick piles of the Cretaceous and Paleogene sediments (Figure 1; [1]) buried a regional forearc basin subducted by the Izanagi/Kula and Pacific Plates. Paleomagnetic studies of the Cretaceous Yezo Supergroup [2,3] showed that the present forearc is divided into some basins developed in different areas. Sedimentary system and forearc basin architecture in the Paleogene was studied in detail by Takano and Waseda [4] and Takano et al. [5].

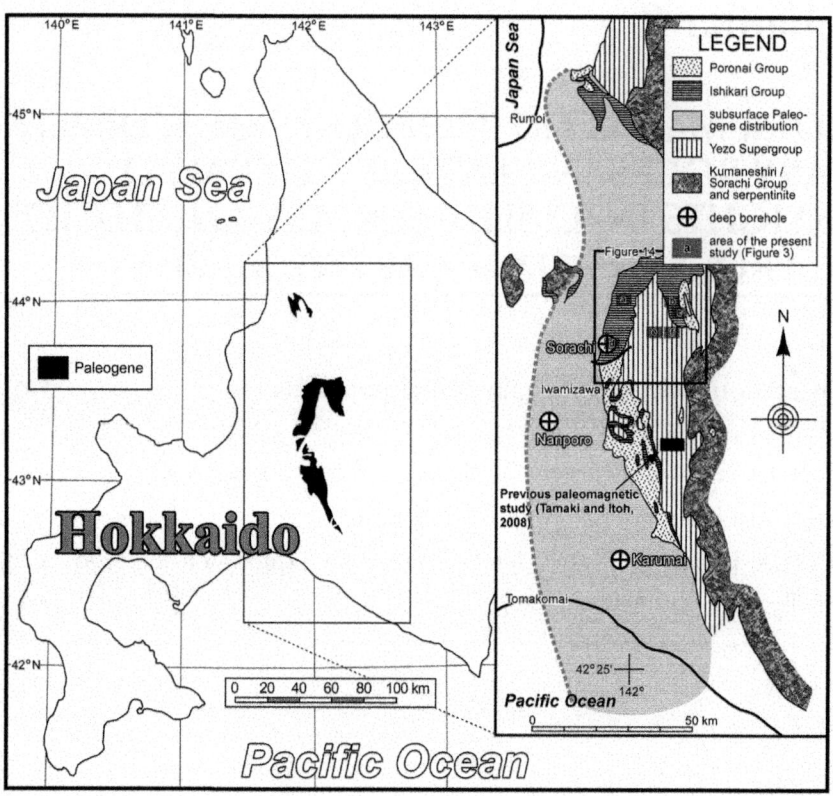

Figure 1. Index map of the study area of the Cretaceous and Paleogene strata. Geologic map is after Editorial Committee of Hokkaido, Regional Geology of Japan [1]

Under the influence of arc-arc collision on the Pacific convergent margin, vigorous mountain building and formation of foreland basins became active since the late Cenozoic. The Ishikari-Teshio belt (see Figure 2) is underlain with thick middle Miocene clastic strata. These are the Kawabata and its correlative formations, derived from the longitudinal mountainous ranges that were uplifted and eroded during that time [6]. It is generally regarded as a typical foreland setting, and the burial history of turbidites and associated coarse clastics of the Kawabata Formation has previously been studied from a sedimentological viewpoint (e.g., [7]). The process through which the Miocene basin developed in central Hokkaido is not only governed by compressive stress in the collision zone, but also by coeval tectonic events like back-arc spreading in the Japan Sea (e.g., [8]) and dextral transcurrent faulting along the Eurasian margin (e.g., [9]).

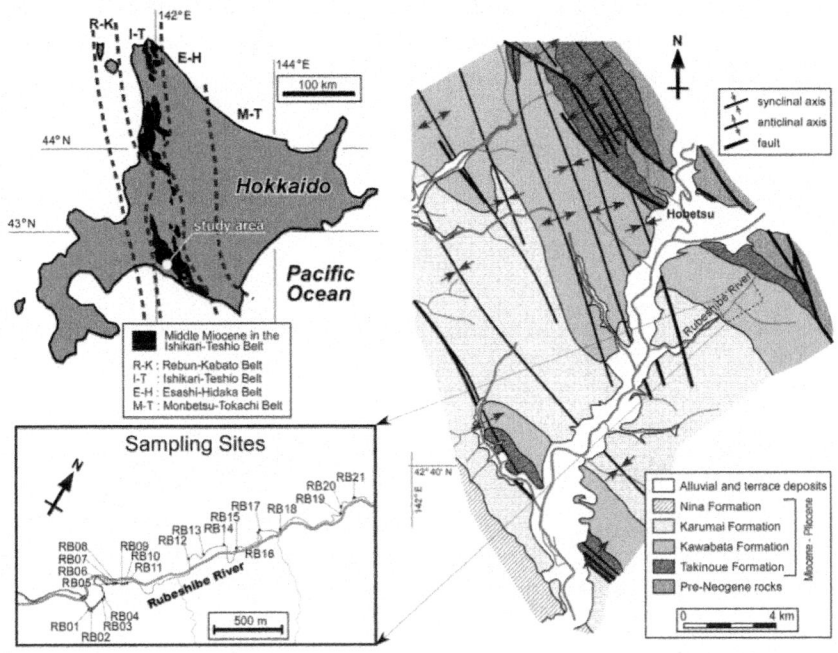

Figure 2. Cenozoic tectonic context of Hokkaido, geology of the study area of the Neogene strata (simplified from Kawakami et al. [7]), and locations of rock magnetic samples.

In this paper, we present preliminary results of rock magnetic analyses of the Cretaceous Yezo Supergroup, the Eocene Ishikari Group and the Miocene Kawabata Formation in order to detect tectonic movements around the basin and to describe the microfabric of sedimentary rocks related to the tectonic regime and sedimentation processes in the mobile zone. This study is an attempt to apply magnetic properties to tectono-sedimentology.

GEOLOGY

Background

The Yezo Supergroup deposited on the Cretaceous forearc and consists of monotonous mudstone intercalated by coarse clastics and ash layers. After a stagnant subsidence stage at the beginning of the Cenozoic, fluvial sediments of the Ishikari Group and its correlative units began to bury depressions on the forearc. As a result of strong deformation and continued sedimentation on the active margin, surface distribution of the Eocene Ishikari Group is rather restricted. However, numerous exploration drilling clarified that voluminous

Paleogene units are concealed under the alluvial plain (Figure 1). Paleogene depositional sequence and facies classification were described by Takano et al. [5]. They are shown in Figure 3 using abbreviations.

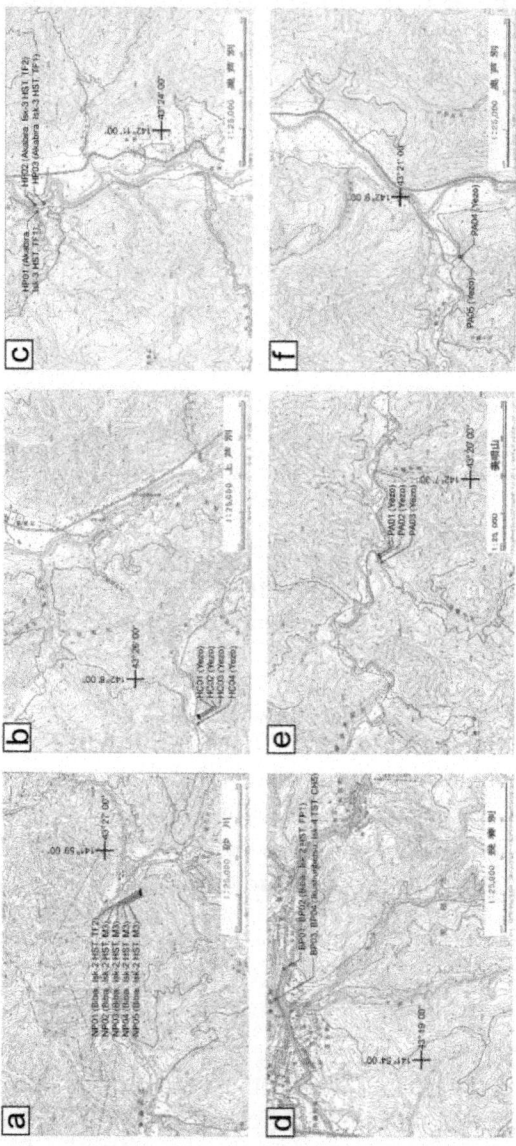

Figure 3. Sampling localities for rock magnetic analyses of the Cretaceous and Paleogene strata. The base maps are parts of the "Sunagawa", "Kamiashibetsu", "Okuashibetsu", "Ikushunbetsu" and "Bibaiyama" 1:25,000 topographic maps published by the Geographical Survey Institute. As for the Paleogene sites (a, c and d), geologic units

(Yezo, Yezo Supergroup; Bibai, Bibai Formation; Akabira, Akabira Formation; Ikush-unbetsu, Ikushunbetsu Formation), depositional sequence and facies classification are shown in parentheses after Takano et al. [5]

The study area of the Kawabata Formation is located in the southern part of the middle Miocene basins of the Ishikari-Teshio belt. Folded sedimentary units are distributed with a NNW-SSE trend, and are cut by numerous faults (Figure 2). The area is divided into the following formations in ascending order [7]: the Takinoue Formation, the Kawabata Formation, the Karumai Formation, and the Nina Formation (Figure 4). They represent the sequence by which an elongate N-S foreland basin was filled. The middle Miocene Kawabata Formation comprises mainly turbidites and associated coarse clastic rocks derived from the eastern hinterland [7].

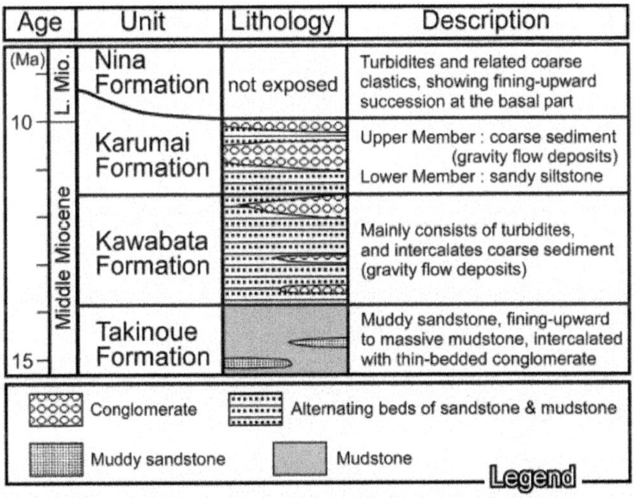

Age	Unit	Lithology	Description
(Ma) L. Mio.	Nina Formation	not exposed	Turbidites and related coarse clastics, showing fining-upward succession at the basal part
10 Middle Miocene	Karumai Formation		Upper Member : coarse sediment (gravity flow deposits) Lower Member : sandy siltstone
	Kawabata Formation		Mainly consists of turbidites, and intercalates coarse sediment (gravity flow deposits)
15	Takinoue Formation		Muddy sandstone, fining-upward to massive mudstone, intercalated with thin-bedded conglomerate

Conglomerate		Alternating beds of sandstone & mudstone	
Muddy sandstone		Mudstone	Legend

Figure 4. Neogene stratigraphy of the study area of the Kawabata Formation

Sedimentary Facies of the Miocene Unit

This study conducted sedimentary facies analysis for the Kawabata Formation along the Rubeshibe River (Figure 2). The analysis revealed that the turbidites of the Kawabata Formation mainly consisted of sheet-flow turbidite facies association and channel-levee facies association (Figure 5). The sheet-flow turbidite facies association comprises aggradational stacking of rhythmic alternating beds of turbidite sandstone and mudstone with rare upward thickening or thinning successions, and is interpreted to be sheet-like turbidites with minor occurrences of depositional lobes, which occupied major part of the trough-like foreland basin fill [7,10]. The channel-levee facies association is composed of thick amalgamated sandstone facies with slump blocks and thinly bedded alternating beds of sandstone and mudstone. These two facies appearing coupled is indicative of an elongated channel-levee system made of the main channel with levees on both sides. These two facies associations are believed to have been deposited in an elongated trough-like foredeep in the foreland basin [7]. The turbidites of the Kawabata Formation commonly contain sedimentary structures indicating paleocurrent directions; e.g., sole marks (mostly flute marks) at the bottom of individual turbidite bed, and current ripples in Bouma Tc division [11].

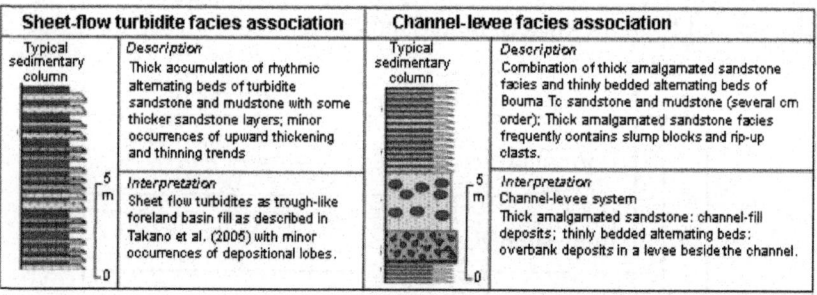

Sheet-flow turbidite facies association		Channel-levee facies association	
Typical sedimentary column	Description Thick accumulation of rhythmic alternating beds of turbidite sandstone and mudstone with some thicker sandstone layers; minor occurrences of upward thickening and thinning trends	Typical sedimentary column	Description Combination of thick amalgamated sandstone facies and thinly bedded alternating beds of Bouma To sandstone and mudstone (several cm order); Thick amalgamated sandstone facies frequently contains slump blocks and rip-up clasts.
	Interpretation Sheet flow turbidites as trough-like foreland basin fill as described in Takano et al. (2005) with minor occurrences of depositional lobes.		Interpretation Channel-levee system Thick amalgamated sandstone: channel-fill deposits; thinly bedded alternating beds: overbank deposits in a levee beside the channel.

Figure 5. Facies association classification of the Kawabata Formation along the Rubeshibe River

ROCK MAGNETISM

We obtained samples for rock magnetic analyses exclusively from fine-grained parts of the target sedimentary units, since fine sedimentary rocks generally preserve stable detrital remanent magnetization (DRM). Few visible markers of the sedimentation process accompany such sediments, so we attempted to measure their microscopic magnetic fabric, which may be related to paleocurrent directions (e.g., [12]).

Basic Measurements

The Cretaceous and Eocene samples were taken from outcrops along the streambed in central Hokkaido (Figure 3) using an engine or electric drill at 21 sites. Samples of the Kawabata Formation were collected with a battery-powered electric drill at 21 sites along the Rubeshibe River (Figure 2). The bedding attitudes were measured on outcrops to allow us to compensate for tectonic tilting later. Between seven and sixteen independently oriented cores

25 mm in diameter were obtained at each site using a magnetic compass. Cylindrical specimens 22 mm in length were cut from each core and the natural remanent magnetization (NRM) of each specimen was measured using a cryogenic magnetometer (model 760-R SRM, 2-G Enterprises). Low-field magnetic susceptibility was measured on a Bartington MS2 susceptibility meter, and the anisotropy of magnetic susceptibility (AMS) was measured using an AGICO KappaBridge KLY-3 S magnetic susceptibility meter. After the basic measurements, pilot specimens with average NRM intensities, directions and susceptibility levels were selected from each site for subsequent demagnetization tests.

Demagnetization Tests

In order to isolate stable components of the remanent magnetization, progressive alternating field demagnetization (PAFD) and progressive thermal demagnetization (PThD) tests were carried out on two pilot specimens per site that had average NRM directions. The PAFD test loading ranged from 0 to 80 mT using a three-axis tumbling system with specimens contained in a μ-metal envelope. The PThD test was performed using an electric furnace, with a residual magnetic field less than 10 nT, beginning at 100 °C and continuing until the specimen was either fully demagnetized and a characteristic remanent magnetization (ChRM) component was isolated, or until the thermal treatment provoked erratic behavior of the magnetic direction. Specimens' low-field bulk magnetic susceptibilities were measured using a susceptibility meter after each PThD step in order to monitor chemical changes in ferromagnetic minerals.

Figure 6 presents typical PThD and PAFD results for the Yezo Supergroup and Ishikari Group. It is obvious that the ChRM direction was not isolated because of unstable behavior in thermal treatment (Figure 6a), overlapping spectra of primary and secondary magnetization (Figure 6b) and partial remagnetization within a site (Figure 6c,d). Therefore further analyses for magnetic granulometry were not applied on the Cretaceous and Eocene samples. On the other hand, PThD treatment was effective for isolating stable ChRM in the sedimentary rocks of the Kawabata Formation. Figure 7 shows typical results of the progressive demagnetization tests.

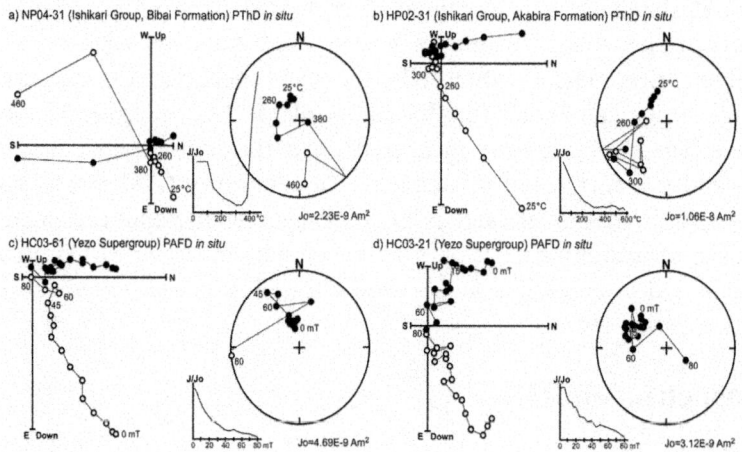

Figure 6. Typical results of progressive thermal demagnetization (PThD) and progressive alternating field demagnetization (PAFD) in geographic coordinates for the Paleogene Ishikari Group (a,b) and the Cretaceous Yezo Supergroup (c,d). On the vector-demagnetization diagrams, solid (open) circles are projection of vector end-points on horizontal (N-S vertical) plane. Equal-area projection and normalized intensity decay curve are shown on the right-side of each vector diagram. Solid (open) circles in equal-area nets are projections on the lower (upper) hemisphere. Numbers attached on data points are demagnetization levels in °C or mT

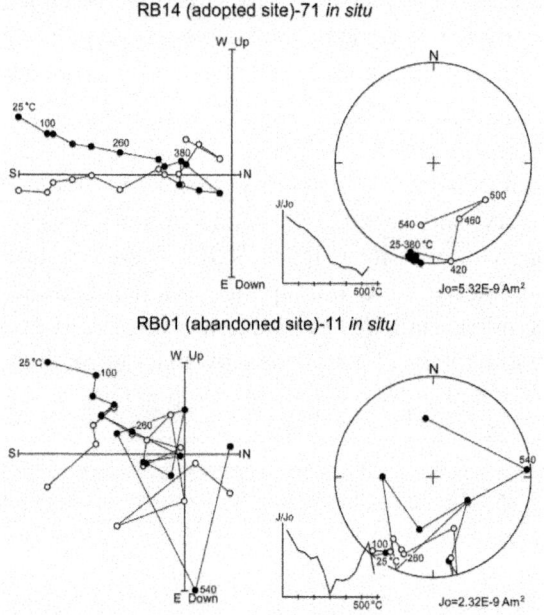

Figure 7. Results of progressive thermal demagnetization for samples of the

Neogene Kawabata Formation with stable (upper) and unstable (lower) magnetization. All coordinates are geographic (*in situ*). Units are bulk remanent intensity. The solid and open circles in the vector-demagnetization diagrams (left) are projections of vector end-points on the horizontal and north-south vertical planes, respectively. The solid and open circles in the equal-area Schmidt nets (right) are projections on the lower and upper hemispheres, respectively.

Hysteresis Properties

Hysteresis parameters were determined for the Kawabata samples with an alternating gradient magnetometer (Princeton Measurements Corporation, MicroMag 2900). Ten sample chips up to 1 mm in size were randomly selected from site RB16, where stable ChRM has been successfully isolated.Figure 8 displays typical hysteresis of the Kawabata mudstones. The raw diagram seems to suggest the absence of ferromagnetic material. After correcting the linear gradient of paramagnetism, a weak ferromagnetic behavior signature can be recognized. Saturation magnetization (Js), saturation remanence (Jrs) and coercive force (Hc) values were determined for all samples from their hysteresis loops. Their relatively low Hc ($\sim$ 100 mT) implies that magnetite is the dominant remanence carrier. After acquiring coercivity of remanence (Hcr) values through backfield demagnetization experiments, we constructed a correlation plot of Jrs/Js versus Hcr/Hc [13] as shown in Figure 9. All the data are plotted in the pseudo-single domain (PSD) region of magnetite.

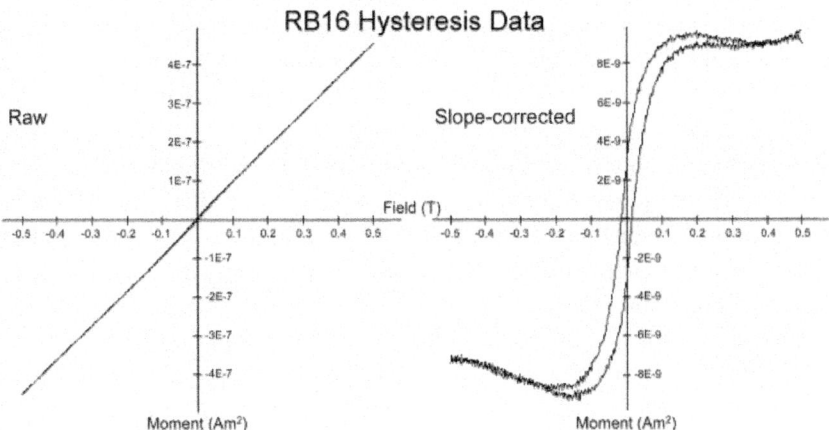

Figure 8. An example of hysteresis loop for a sample of the Kawabata Formation from site RB16 (Left: raw data, Right: data corrected for slope of paramagnetism).

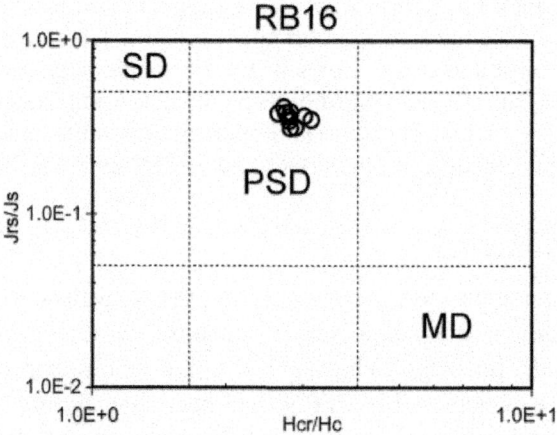

Figure 9. Logarithmic plot of hysteresis parameters [13] of ten samples of the Kawabata Formation from site RB16. Abbreviations: SD, single domain; PSD, pseudo-single domain; MD, multi-domain.

DISCUSSION

Rotational Motions

We found stable magnetic components at three sites of the Kawabata Formation. Their directions were determined with a three-dimensional least squares analysis technique [14]. Figure 10 and Table 1 present site-mean ChRM directions obtained from the Kawabata Formation. They exhibit antipodal directions, and precision parameter (κ) improves after tilt correction. Although the number of data points is minimal for tectonic discussion, we can interpret the site-mean directions as a record of the Earth's dipole magnetic field, acquired before the strata tilted. The declination of the formation mean exhibits a significant westerly deflection, which suggests counterclockwise rotation of the study area.

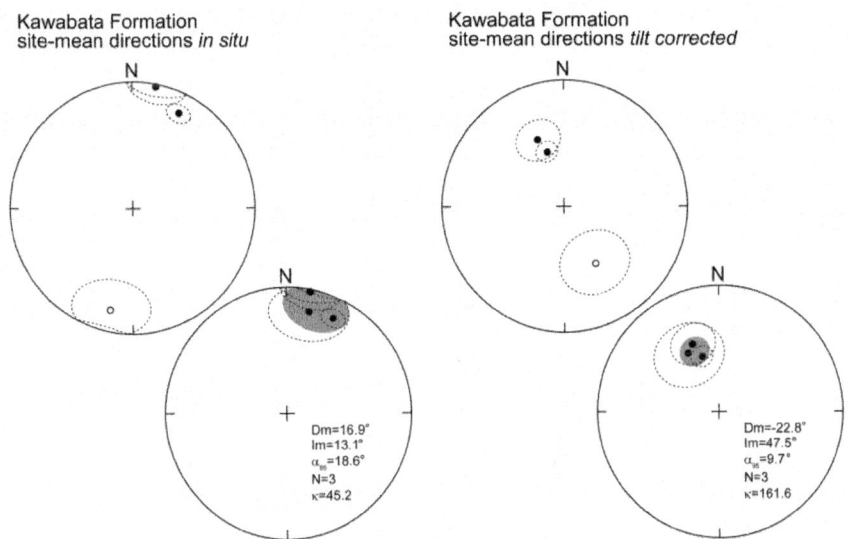

Figure 10. Site-mean ChRM directions of the Kawabata Formation in the study area. The solid and open circles in all the equal-area nets are projections on the lower and upper hemispheres, respectively. Dotted ovals show 95 % confidence limits. Lower diagrams are polarity-converted for calculating formation mean directions and Fisher's precision parameters as annotated in the diagrams (Shaded ovals depict 95 % confidence for the formation means).

Table 1. Paleomagnetic directions of the Kawabata Formation

Site	Latitude	Longitude	D	I	Dc	Ic	α95	κ	N	φ	λ
RB14	42.7361	142.1771	-167.1	-18.7	151.2	-47.0	21.9	13.1	5	62.5	29.5
RB16	42.7379	142.1793	11.4	2.8	-21.8	42.5	14.0	14.4	9	64.5	13.9
RB17	42.7381	142.1793	26.8	17.4	-17.4	52.7	6.8	66.5	8	73.3	23.1

[i] - D and I, *in situ* site-mean declination and inclination before tilt correction in degrees, respectively; Dc and Ic, site-mean declination and inclination after tilt correction in degrees, respectively; α95, radius of 95% confidence circle in degrees; κ, precision parameter; N, number of specimens; φ and λ, latitude (N) and longitude (E) of north-seeking virtual geomagnetic pole for untilted site-mean direction in degrees, respectively.

A previous study [15] suggested a clockwise tectonic rotation around central Hokkaido based on a paleomagnetic study of the Kawabata Formation. Takeuchi et al. [16] proposed a coherent rotational model with 'domino-style' rigid crustal blocks. However, Tamaki et al. [17] criticized the block rotation scheme as being overly simplistic based on differential rotations inferred from Oligocene paleomagnetic data. They restored crustal deformation in central Hokkaido using dislocation modeling, and found complicated vertical-axis rotations around terminations of the faults that contributed to the formation of N-S elongate sedimentary basins. Figure 11 demonstrates differential rotation in central Hokkaido since the middle Miocene.

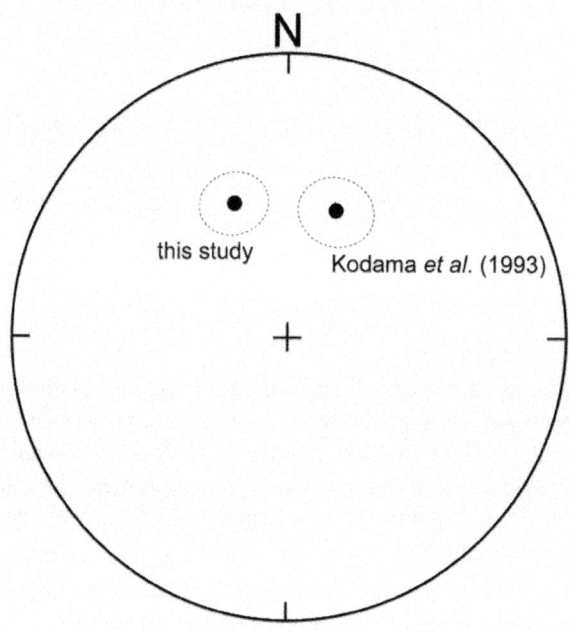

Figure 11. Comparison of the mean paleomagnetic directions of the Kawabata Formation in central Hokkaido between this study and [15]. Data are plotted on the lower hemisphere of the equal-area projection. Dotted ovals represent 95 % confidence limits.

Sedimentation Process Inferred From AMS Fabric

We found that the AMS fabric (orientation of principal axes) were precisely determined at all the sampled localities. Tables 2 and 3 show the AMS parameters for the Cretaceous/Eocene units and the Miocene unit, respectively. Figure 12 delineates typical AMS fabric obtained from the Ishikari (left) and Yezo (right) samples. After tilt-correction, the maximum (K_1) and intermediate (K_2) axes of AMS are bound to the horizontal plane with a subtle imbrication suggestive of hydrodynamic forcing.

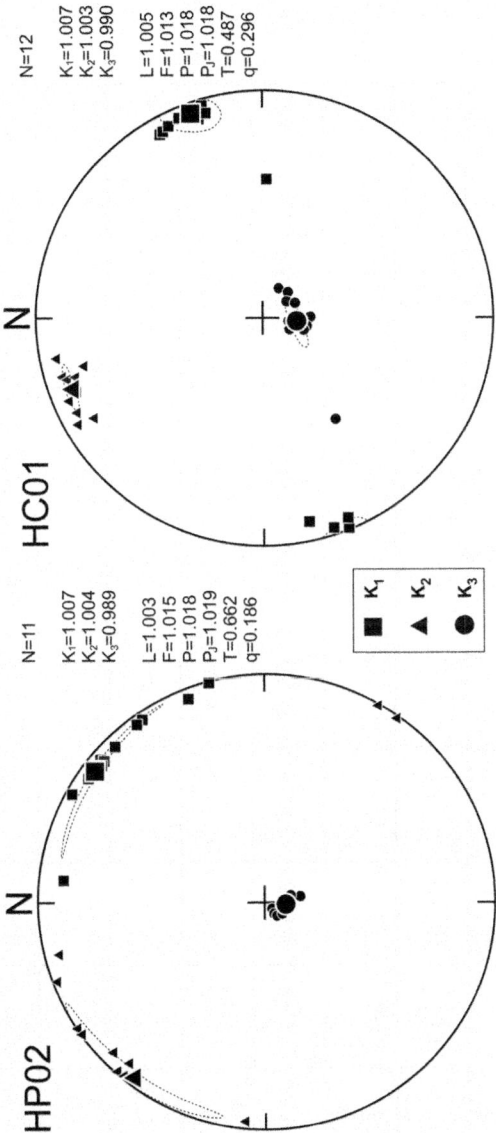

Figure 12. Anisotropy of magnetic susceptibility (AMS) fabric (principal susceptibility axes) for all specimens of typical sites of the Ishikari Group (HP02) and Yezo
Supergroup (HC01) plotted on the lower hemisphere of equal-area projections. Data
are shown in stratigraphic coordinates. Ovals surrounding mean directions of three
axes (shown by larger symbols) are 95% confidence regions. See Table 2 for all the
AMS parameters.

Table 2. Site-mean AMS parameters of the Paleogene and Cretaceous units in central Hokkaido

Site	N	K_1Str.	K_2Str.	K_3Str.	L	F	P	P'	T	q	Unit / Sequence
		(D, I)	(D, I)	(D, I)	(K_1/K_2)	(K_2/K_3)	(K_1/K_3)				
Paleogene											
BP01	11	1.010	1.005	0.984	1.005	1.021	1.026	1.028	0.619	0.213	Bb / Isk-2HST
		(207, 4)	(297, 1)	(35, 86)							
BP02	11	1.009	1.008	0.982	1.001	1.026	1.027	1.031	0.917	0.043	Bb / Isk-2HST
		(226, 3)	(136, 4)	(0, 85)							
BP03	11	1.010	1.005	0.985	1.005	1.020	1.025	1.027	0.593	0.229	Ik / Isk-4TST
		(233, 9)	(141, 7)	(16, 78)							
BP04	13	1.010	1.002	0.988	1.008	1.014	1.022	1.022	0.259	0.458	Ik / Isk-4TST
		(225, 6)	(134, 3)	(15, 83)							
HP01	18	1.010	1.004	0.987	1.006	1.017	1.023	1.024	0.479	0.303	Ak / Isk-3HST

Sample	N	Dir 1	k_1	Dir 2	k_2	Dir 3	k_3	L	F	P	P'	T	q	Lithology
HP02	11	(271, 4)	1.007	(180, 8)	1.004	(29, 81)	0.989	1.003	1.015	1.018	1.019	0.662	0.186	Ak / Isk-3HST
		(38, 6)		(307, 4)		(187, 83)								
NP01	15	(264, 2)	1.009	(174, 7)	1.006	(11, 83)	0.985	1.003	1.022	1.025	1.027	0.762	0.128	Bb / Isk-2HST
NP02	9	(252, 10)	1.013	(162, 2)	1.006	(59, 80)	0.981	1.006	1.026	1.032	1.034	0.597	0.227	Bb / Isk-2HST
NP03	10	(253, 1)	1.006	(163, 7)	1.004	(349, 83)	0.989	1.002	1.016	1.018	1.019	0.779	0.118	Bb / Isk-2HST
NP04	7	(80, 10)	1.009	(170, 4)	1.006	(282, 79)	0.985	1.002	1.022	1.024	1.027	0.791	0.111	Bb / Isk-2HST
NP05	9		1.007		1.003		0.990	1.004	1.013	1.017	1.018	0.536	0.264	Bb / Isk-2HST

	N	(39, 28)	(141, 21)	(261, 54)	Cretaceous					
HC01	12	1.007	1.003	0.990	1.005	1.013	1.018	1.018	0.487	0.296
		(71, 5)	(340, 11)	(187, 78)						
HC02	10	1.011	1.005	0.984	1.006	1.021	1.027	1.028	0.541	0.262
		(66, 2)	(156, 4)	(315, 86)						
HC03	11	1.010	1.002	0.988	1.008	1.014	1.022	1.022	0.274	0.447
		(51, 4)	(321, 11)	(163, 78)						
PA01	13	1.006	1.003	0.991	1.003	1.012	1.015	1.016	0.650	0.193
		(336, 11)	(70, 19)	(217, 68)						
PA04	14	1.017	1.014	0.969	1.004	1.046	1.049	1.055	0.852	0.078
		(8, 9)	(99, 11)	(240, 76)						

[i] - *N* denotes the number of specimens. Directions of AMS principal axes are in stratigraphic coordinates. Abbreviations for the Paleogene geologic units: Ak, Akabira Formation; Bb, Bibai Formation; Ik, Ikushunbetsu Formation. Depositional sequence is after Takano & Waseda [4] and Takano et al. [5].

Figure 13 delineates typical AMS fabrics of the Kawabata Formation. Site RB08 typifies an elongate (prolate) fabric reflecting aligned detrital grains. Site RB14 has highly oblate fabric, as shown by a positive T parameter near unity. This fabric is essentially confined to the bedding plane under gravitational force. As the hysteresis study showed a negligible amount of ferromagnetic material in the Kawabata samples (Figure 8), we consider the AMS fabric as being governed simply by the shape anisotropy of paramagnetic minerals, i.e. alignments of elongate or platy grains such as amphibole or mica.

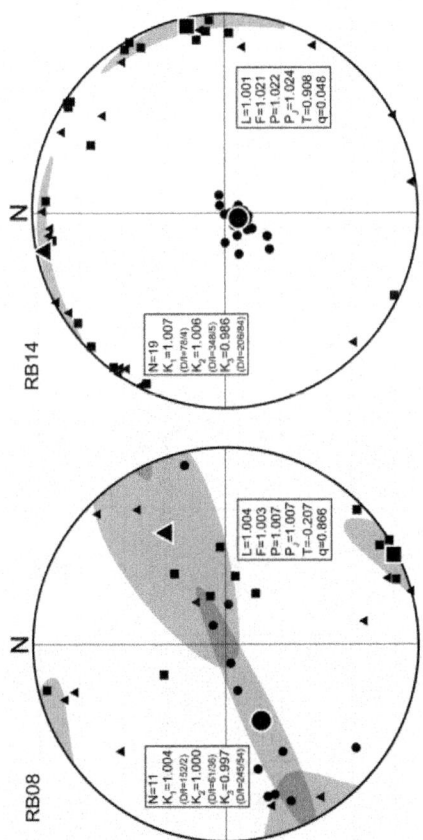

Figure 13. Typical tilt-corrected AMS fabric for the Kawabata Formation muddy samples. Prolate (left) and oblate (right) fabrics are numerically described by negative and positive T parameters, respectively, posted on the equal-area diagrams. All the data are plotted on the lower hemisphere. Square, triangular and circular symbols represent orthogonal maximum (K_1), intermediate (K_2), and minimum (K_3) AMS principal axes, respectively, and larger symbols show their mean directions. Shaded areas are 95 % confidence limits based upon Bingham statistics.

Table 3. Site-mean AMS parameters of the Kawabata Formation

Site	N	K_1 (D, I)	K_2 (D, I)	K_3 (D, I)	L (K_1/K_2)	F (K_2/K_3)	P (K_1/K_3)	P'	T	q
RB01	11	1.013 (16, 6)	1.006 (107, 7)	0.981 (242, 81)	1.007	1.025	1.033	1.034	0.546	0.260
RB02	11	1.007 (3, 20)	1.003 (98, 11)	0.991 (215, 66)	1.004	1.012	1.016	1.017	0.508	0.282
RB03	12	1.006 (326, 11)	1.002 (58, 11)	0.992 (190, 74)	1.004	1.010	1.014	1.015	0.475	0.304
RB04	13	1.008 (159, 2)	1.002 (249, 6)	0.990 (52, 83)	1.005	1.013	1.018	1.019	0.406	0.352
RB05	15	1.005 (4, 19)	1.001 (101, 19)	0.993 (232, 62)	1.004	1.008	1.012	1.012	0.330	0.404
RB06	14	1.007 (336, 5)	1.001 (66, 7)	0.992 (208, 82)	1.005	1.009	1.014	1.015	0.256	0.459
RB07	12	1.005	0.999	0.996	1.007	1.002	1.009	1.009	-0.459	1.151

Sample	N										
RB08	11	1.004	1.000	0.997	1.004	1.003	1.007	1.007	-0.207	0.866
		(135, 24)	(41, 9)	(291, 64)						
RB09	17	1.005	1.001	0.994	1.004	1.007	1.011	1.011	0.313	0.416
		(152, 2)	(61, 36)	(245, 54)						
RB10	10	1.010	1.007	0.983	1.003	1.024	1.027	1.030	0.777	0.120
		(121, 1)	(31, 7)	(218, 83)						
RB11	12	1.002	0.999	0.998	1.003	1.001	1.004	1.004	-0.410	1.090
		(343, 2)	(252, 12)	(84, 78)						
RB12	12	1.005	1.004	0.991	1.002	1.013	1.015	1.016	0.763	0.127
		(313, 42)	(105, 45)	(210, 15)						
RB13	9	1.004	1.002	0.994	1.001	1.009	1.010	1.011	0.732	0.144
		(3, 6)	(93, 1)	(192, 84)						
RB14	19	1.007	1.006	0.986	1.001	1.021	1.022	1.024	0.908	0.048
		(119, 26)	(214, 11)	(325, 61)						
		(78, 4)	(348, 5)	(206, 84)						

	N									
RB15	12	1.009 (188, 10)	1.004 (96, 10)	0.986 (324, 76)	1.005	1.018	1.023	1.024	0.558	0.251
RB16	15	1.013 (281, 4)	1.010 (11, 5)	0.977 (152, 83)	1.003	1.034	1.036	1.041	0.848	0.080
RB17	14	1.009 (292, 6)	1.003 (201, 6)	0.987 (66, 81)	1.006	1.016	1.022	1.023	0.455	0.318
RB18	7	1.009 (120, 1)	1.001 (30, 2)	0.990 (234, 87)	1.008	1.011	1.018	1.018	0.169	0.528
RB19	13	1.012 (26, 15)	1.003 (277, 50)	0.985 (127, 36)	1.009	1.018	1.028	1.028	0.324	0.411
RB20	11	1.013 (300, 19)	1.005 (205, 15)	0.982 (78, 66)	1.008	1.023	1.031	1.032	0.458	0.317
RB21	17	1.015 (215, 75)	1.005 (90, 9)	0.980 (358, 12)	1.010	1.026	1.036	1.037	0.433	0.335

[i] - N is the number of specimens. Directions of AMS principal axes are in stratigraphic coordinates.

Sedimentological context of the AMS fabric is demonstrated in Figures 14 and 15. Paleocurrent directions inferred from the Eocene AMS data tend to align in N-S azimuth (Figure 14), and accord with development process of the forearc basin [4]. Takano and Waseda [4] demonstrated that the Eocene paleo-Ishikari basin experienced differential subsidence during deposition. Such deformation may be related to longstanding strike-slip faulting around central Hokkaido [17], and tectono- / sedimentological context of the AMS fabric will be better evaluated in the light of quantitative study of basin-forming processes described in this book. For reliable interpretation of AMS data, it is necessary to assess properties of ferromagnetic minerals, such as composition, grain size and contribution to bulk magnetic susceptibility, as shown in this paper.

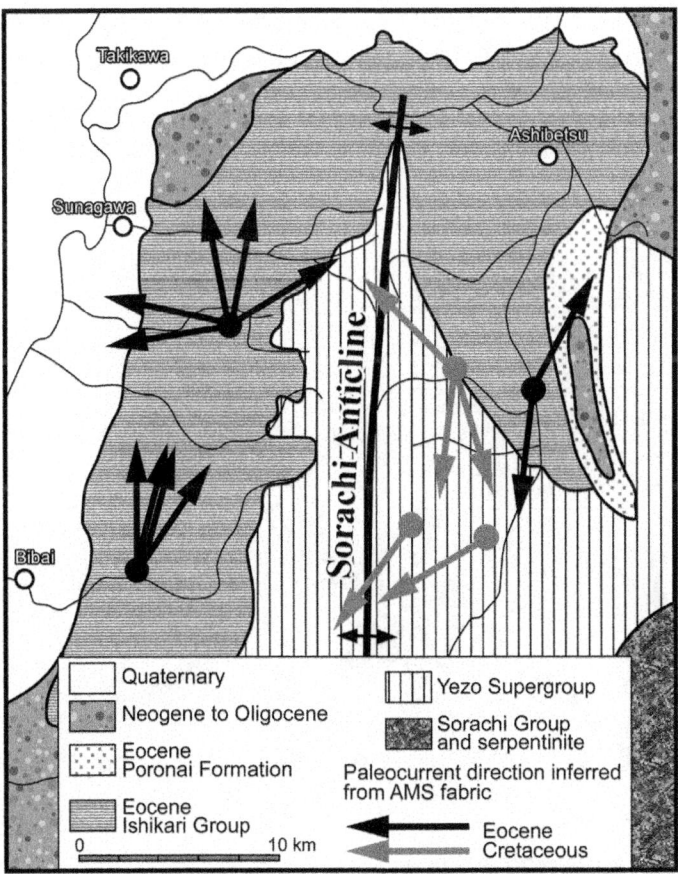

Figure 14. Paleocurrent directions inferred from AMS fabric of the Paleogene and Cretaceous samples. Geologic map is compiled from Editorial Committee of Hokkaido, Regional Geology of Japan [1] and Takano and Waseda [4].

Our field survey revealed indicators of paleocurrent directions in the Kawabata Formation along the Rubeshibe River as depicted in Figure 15. After correction for the counterclockwise rotation identified in our paleomagnetic study, most of the markers indicate a westward current direction with minor southward flow contributions. This is consistent with a tectono-sedimentary model of rapid burial of the Miocene N-S foreland basin by clastics derived from the eastern collision front presented in such research as Kawakami et al. [7]. Notably, the imbrication of the oblate AMS fabric matches visible sedimentary structures. Although the transport direction of muddy detrital material spilled out of a levee is not necessarily parallel to the turbidity current within a channel, AMS data can serve to indicate paleocurrents after the contributors to the magnetic fabric have been identified. Also note that K_1 of prolate samples (with negative T parameters) tend to align perpendicular to the paleocurrent direction, implying that elongate grains roll on the sediment surface.

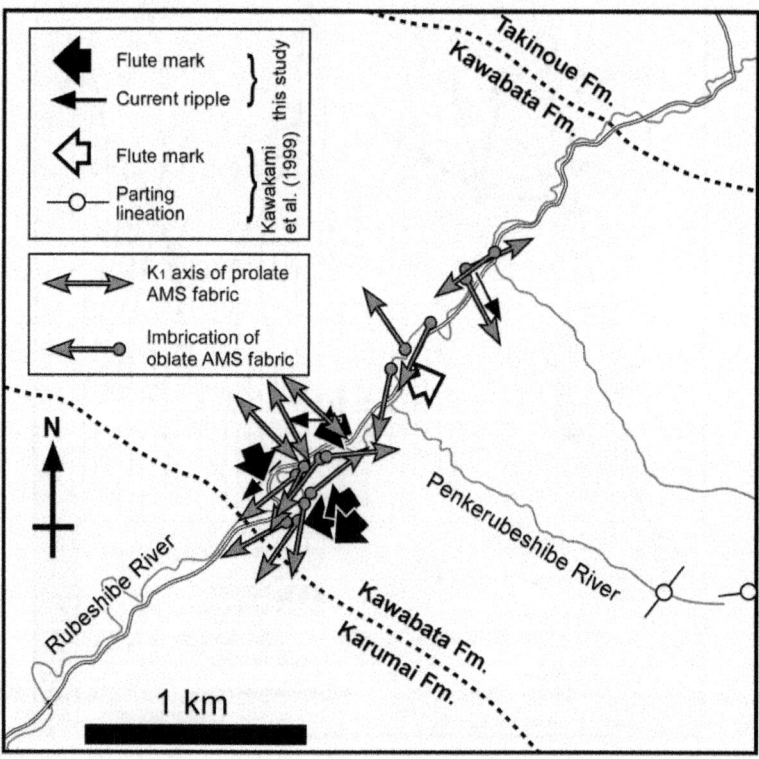

Figure 15. Paleocurrent map of the Kawabata Formation around the Rubeshibe River route. Formation boundaries are after Kawakami et al. [7].

Figure 16 delineates groups of microscopic fabrics identified in the Kawabata Formation as a function of the AMS shape parameter (T). The intensity of alignment forcing inferred from AMS data is closely related to sedimentary facies (shown on the right in the figure) determined by field observation. For example, weak hydrodynamic forcing corresponds to fine rhythmically alternating facies in channel-levee systems. Thus, the sedimentological context of muddy sediments' AMS fabric can be interpreted in the light of sandy sediments' facies analysis.

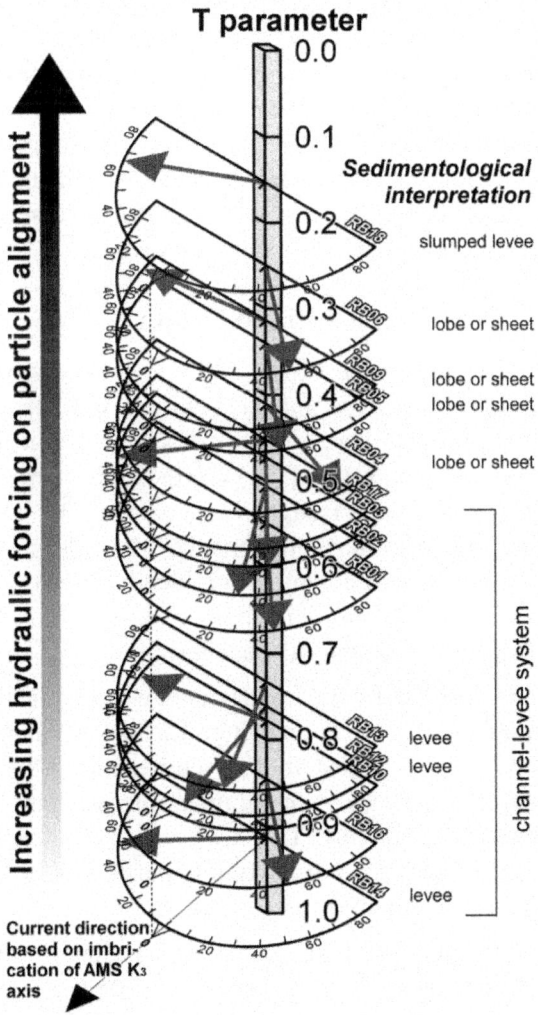

Figure 16. AMS paleocurrent indicators of the Kawabata Formation. Directions of K_1 (gray arrows) are shown as acute angles from the dotted baseline of K_3 axis imbrica-

tion. Vertical positions of the data are based on the T parameter. Samples with negative T values are excluded from the diagram because such cases have a large scatter in the K_3 directions.

Azimuths of AMS maxima in natural sediments vary significantly, reflecting the size or shape of magnetic grains and changes in current velocities (e.g., [18]). Figure 16 presents the relationship between paleocurrent proxies estimated from the imbrication of the AMS minimum axis (K_3) and the K_1 trend. Tarling and Hrouda [19] stated that the angle between K_3 and K_1 changes as a function of current velocity and the slope of the sedimentary surface. Our result suggests that the orientation between those AMS sedimentary indicators can vary, regardless of the level of hydraulic forcing, based on the shape parameter (T), which implies development of a preferred orientation. Although the AMS fabric is a diagnostic tool for patterns of sediment transportation, laboratory-based experiments that analyze natural sediments under conditions where a few of the prevailing factors are controlled, are essential to allow firm sedimentological interpretation of formation processes.

Re-Deposition Experiment and the Origin of AMS

In order to consider the origin of the AMS in the Kawabata samples, we organized a re-deposition experiment. A silty sandstone (SP1C-1) and a mudstone (SP2F-1) samples were crushed and sieved into coarse, medium and fine fractions. The fine fraction (< 63 μm) was then separated into magnetic and non-magnetic fractions with an isodynamic separator. The 'magnetic' fraction actually contained no ferromagnetic opaque minerals such as magnetite, but had abundant biotite and common hornblende. It also contained garnet, probably derived from metamorphic rocks exposed around the hinterlands during the rapid deposition of the Miocene turbidite.

A suspension of the fine fraction was poured into a vertically settled plastic tube 1 m in length and 2.5 cm in diameter, filled with water. This deposit of artificial sediment was dehydrated at room temperature. After being soaked in an adhesive resin, the samples were trimmed into standard-sized specimens for rock-magnetic measurements. The AMS was measured with an AGICO KappaBridge KLY-3 S magnetic susceptibility meter. The AMS parameters for the artificial samples are summarized in Table 4.

Figure 17 presents the magnitudes of magnetic fabrics in natural sedimentary rocks and the re-deposited sediments of the Kawabata Formation. Obviously, the magnetic separation results in remarkable decrease of both the bulk susceptibility and the degree of anisotropy (P_j). It is also noteworthy that the shape parameter (T) of the artificial sediments is almost null, suggesting a neutral magnetic fabric. The directions of the principal AMS axes (see Table

4) are not bound to the horizontal plane or to geomagnetic north. Thus, the detrital particles, free from paramagnetic minerals having shape anisotropy, like platy biotite, are deposited without any gravitational or geomagnetic forcing, creating an isotropic sediment.

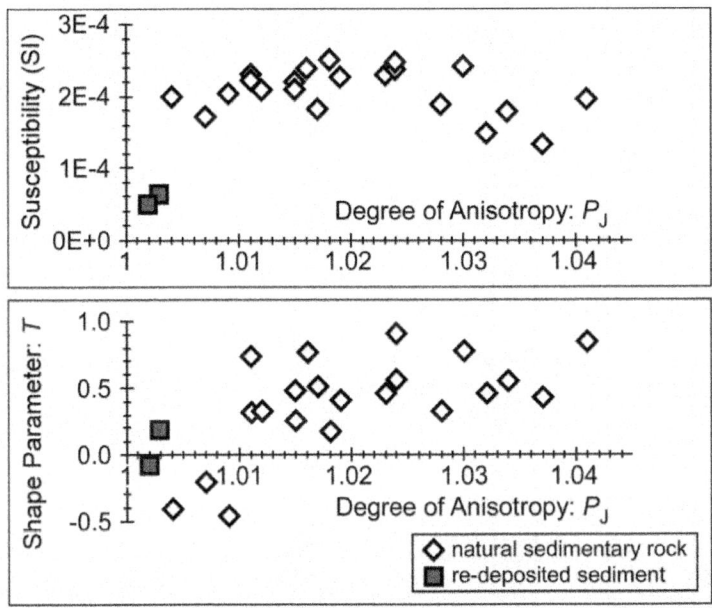

Figure 17. Magnitudes of magnetic fabrics in natural samples and re-deposited non-magnetic fine particles of the Kawabata Formation.

Table 4. AMS parameters of re-deposited non-magnetic fine fraction of the Kawabata Formation

Sample	N	K^1	K^2	K^3	$L (K_1/K_2)$	$F (K_2/K_3)$	$P (K_1/K_3)$	P'	T	q
		(D, I)	**(D, I)**	**(D, I)**						
SP1C-1	1	1.0009	1.0000	0.9992	1.001	1.001	1.002	1.002	-0.080	0.740
		(167, 75)	(265, 2)	(356, 15)						
SP2F-1	1	1.0014	1.0002	0.9984	1.001	1.002	1.003	1.003	0.180	0.517
		(250, 27)	(343, 6)	(84, 62)						

[i] - N is the number of specimens. Directions of principal axes of AMS are shown in *in situ* coordinates.

SUMMARY

Rock-magnetic investigation of sedimentary rocks provides insights into the basin's formation and sedimentation processes on an active margin. Cretaceous (Yezo Supergroup) ~ Eocene (Ishikari Group) strata and middle Miocene (Kawabata Formation) turbidites in central Hokkaido represent forearc and foreland settings, respectively. Progressive demagnetization successfully isolated characteristic remanent magnetization (ChRM) of the Kawabata Formation. Mean declination of the formation's ChRM exhibited significant westerly deflection, suggesting counterclockwise rotation of the study area since the middle Miocene. This differs from previous reports that indicated clockwise rotation. We attribute the difference to complicated deformation around the terminations of faults that form the N-S elongate Kawabata sedimentary basin. Anisotropy of magnetic susceptibility (AMS) principal axes were clearly determined for both the Cretaceous/Paleogene samples and Neogene samples, and regarded as a proxy of sediment influx directions. Paleocurrent directions inferred from the Eocene AMS data tend to align in N-S azimuth (Figure 14), and accord with the results of sedimentological paleoenvironment reconstruction, which suggest a northward downstream trend in fluvial to tidal estuarine systems [4]. As for the Cretaceous, further acquisition of AMS data is necessary to assess the effect of intensive syn-depositional deformation of the forearc [20]. After correcting for the tectonic rotation, most of the paleocurrent markers in the Kawabata Formation indicated a westward current direction with minor southward flow contributions, consistent with a sedimentary model that envisions burial of the Miocene N-S foreland basin by clastics derived from the eastern collision front. The intensity of alignment forcing of sedimentary particles inferred from the shape parameter (T) of the AMS data was closely related to sedimentary facies observed in the field. In investigating the origin of the AMS fabrics of turbidite deposits of the Kawabata Formation, we conducted a re-deposition experiment of fine detrital particles with no magnetic fraction including paramagnetic minerals with relatively high magnetic susceptibility, which demonstrated the significance of the alignment of paramagnetic minerals having shape anisotropy.

ACKNOWLEDGEMENTS

The authors are grateful to N. Ishikawa for the use of the rock-magnetic laboratory at Kyoto University and for thoughtful suggestions in the course of the magnetic analyses. We thank S. Oshimbe and S. Nishizaki for their help with field work. Thanks are also due to N. Yamashita and Y. Danhara for their support in mineral separation. Constructive review comments by G. Kawakami greatly helped to improve early version of the manuscript.

REFERENCES

1.   Editorial Committee of Hokkaido, Regional Geology of Japan. Regional Geology of Japan, Part 1: Hokkaido. Tokyo: Kyoritsu Shuppan; 1990.

2.   Tamaki M, Itoh Y. Tectonic implications of paleomagnetic data from upper Cretaceous sediments in the Oyubari area, central Hokkaido, Japan. Island Arc 2008; 17: 270-284.

3.   Tamaki M, Oshimbe S, Itoh Y. A large latitudinal displacement of a part of Cretaceous forearc basin in Hokkaido, Japan: paleomagnetism of the Yezo Supergroup in the Urakawa area. Journal of Geological Society of Japan 2008; 114: 207-217.

4.   Takano O, Waseda A. Sequence stratigraphic architecture of a differentially subsiding bay to fluvial basin: the Eocene Ishikari Group, Ishikari Coal Field, Hokkaido, Japan. Sedimentary Geology 2003; 160: 131-158.

5.   Takano O, Waseda A, Nishita H, Ichinoseki T, Yokoi K. Fluvial to bay-estuarine system and depositional sequences of the Eocene Ishikari Group, central Hokkaido. Journal of Sedimentological Society of Japan 1998; 47: 33-53.

6.   Miyasaka S, Hoyanagi K, Watanabe Y, Matsui M. Late Cenozoic mountain-building history in central Hokkaido deduced from the composition of conglomerate. Monograph of the Association for the Geological Collaboration in Japan 1986; 31: 285-294.

7.   Kawakami G, Yoshida K, Usuki T. Preliminary study for the Middle Miocene Kawabata Formation, Hobetsu district, central Hokkaido, Japan: special reference to the sedimentary system and the provenance. Journal of the Geological Society of Japan 1999; 105: 673-686.

8.   Otofuji Y, Kambara A, Matsuda T, Nohda S. Counterclockwise rotation of northeast Japan: paleomagnetic evidence for regional extent and timing of rotation. Earth and Planetary Science Letters 1994; 121: 503-518.

9.   Itoh Y, Tsuru T. Evolution history of the Hidaka-oki (offshore Hidaka) basin in the southern central Hokkaido, as revealed by seismic interpretation, and related tectonic events in an adjacent collision zone. Physics of the Earth and Planetary Interiors 2005; 153: 220-226.

10.   Takano O, Tateishi M, Endo M. Tectonic controls of a backarc trough-fill turbidite system; the Pliocene Tamugigawa Formation in the Niigata-Shin'etsu inverted rift basin, Northern Fossa Magna, central Japan. Sedimentary Geology 2005; 176: 247-279.

11.   Bouma AH. Sedimentology of Some Flysch Deposits; A Graphic

Approach to Facies Interpretation. Amsterdam: Elsevier; 1962.

12. Kawamura K, Ikehara K, Kanamatsu T, Fujioka K. Paleocurrent analysis of turbidites in Parece Vela Basin using anisotropy of magnetic susceptibility. Journal of the Geological Society of Japan 2002; 108: 207-218.

13. Day R, Fuller M, Schmidt VA. Hysteresis properties of titanomagnetites: grain-size and compositional dependence. Physics of Earth and Planetary Interiors 1977; 13: 260-267.

14. Kirschvink JL. The least-squares line and plane and the analysis of palaeomagnetic data. Geophysical Journal of the Royal Astronomical Society 1980; 62: 699-718.

15. Kodama K, Takeuchi T, Ozawa T. Clockwise tectonic rotation of Tertiary sedimentary basins in central Hokkaido, northern Japan. Geology 1993; 21: 431-434.

16. Takeuchi T, Kodama K, Ozawa T. Paleomagnetic evidence for block rotations in central Hokkaido-south Sakhalin, Northeast Asia. Earth and Planetary Science Letters 1999; 169: 7-21.

17. Tamaki M, Kusumoto S, Itoh Y. Formation and deformation processes of late Paleogene sedimentary basins in southern central Hokkaido, Japan; paleomagnetic and numerical modeling approach. Island Arc 2010; 19: 243-258.

18. Ledbetter MT, Ellwood BB. Spatial and temporal changes in bottom-water velocity and direction from analysis of particle size and alignment in deep-sea sediment. Marine Geology 1980; 38: 245-261.

19. Tarling DH, Hrouda F. The Magnetic Anisotropy of Rocks. London: Chapman & Hall; 1993.

20. Tamaki M, Tsuchida K, Itoh Y. Geochemical modeling of sedimentary rocks in the central Hokkaido, Japan: Episodic deformation and subsequent confined basin-formation along the eastern Eurasian margin since the Cretaceous. Journal of Asian Earth Sciences 2009; 34: 198-208.

CITATION

CHAPTER 1

Al-Maamori, H., El Naggar, M. and Micic, S. (2014) A Compilation of the Geo-Mechanical Properties of Rocks in Southern Ontario and the Neighbouring Regions. Open Journal of Geology, 4, 210-227. doi:10.4236/ojg.2014.45017.

CHAPTER 2

Lixin Wu and Shanjun Liu (2009). Remote Sensing Rock Mechanics and Earthquake Thermal Infrared Anomalies, Advances in Geoscience and Remote Sensing, Gary Jedlovec (Ed.), ISBN: 978-953-307-005-6, InTech, DOI: 10.5772/8292.

CHAPTER 3

X. Pan, Z. Feng, G. Dai and H. Liu, "Roughness Research of Center Profile Curve on Rock Fracture Surface Based on Statistical Method," Geomaterials, Vol. 3 No. 2, 2013, pp. 47-53. doi: 10.4236/gm.2013.32006.

CHAPTER 4

A. Bery and R. Saad, "Correlation of Seismic P-Wave Velocities with Engineering Parameters (N Value and Rock Quality) for Tropical Environmental Study," International Journal of Geosciences, Vol. 3 No. 4, 2012, pp. 749-757. doi: 10.4236/ijg.2012.34075.

CHAPTER 5

Yifeng Chen and Chuangbing Zhou (2011). Stress/Strain-Dependent Properties of Hydraulic Conductivity for Fractured Rocks, Developments in Hydraulic Conductivity Research, Dr. Oagile Dikinya (Ed.), ISBN: 978-953-307-470-2, InTech, DOI: 10.5772/16007.

CHAPTER 6

Hu-Dan Tang, Zhen-De Zhu, Ming-Li Zhu, and Heng-Xing Lin, "Mechanical Behavior of 3D Crack Growth in Transparent Rock-Like Material Containing Preexisting Flaws under Compression," Advances in Materials Science and Engineering, vol. 2015, Article ID 193721, 10 pages, 2015. doi:10.1155/2015/193721

CHAPTER 7

Yijiang Peng, Qing Guo, Zhaofeng Zhang, and Yanyan Shan, "Application of Base Force Element Method on Complementary Energy Principle to Rock Mechanics Problems," Mathematical Problems in Engineering, vol. 2015, Article ID 292809, 16 pages, 2015. doi:10.1155/2015/292809

CHAPTER 8

Yasuto Itoh, Machiko Tamaki and Osamu Takano (2013). Rock Magnetic Properties of Sedimentary Rocks in Central Hokkaido — Insights into Sedimentary and Tectonic Processes on an Active Margin, Mechanism of Sedimentary Basin Formation - Multidisciplinary Approach on Active Plate Margins, Dr. Yasuto Itoh (Ed.), ISBN: 978-953-51-1193-1, InTech, DOI: 10.5772/56650.

INDEX